A ~~SPIRITUAL~~ SCIENTIFIC APPROACH TO

REWIRING SELF LOVE

WHERE ANCIENT WISDOM MEETS NEUROSCIENCE:
UNDERSTAND YOUR PATTERNS, RELEASE YOUR BLOCKS, AND CREATE LASTING SELF-WORTH

FIONA SOUTTER

© COPYRIGHT 2025 - ALL RIGHTS RESERVED.

The content contained within this book may not be reproduced, duplicated or transmitted without direct written permission from the author or the publisher.

Under no circumstances will any blame or legal responsibility be held against the publisher, or author, for any damages, reparation, or monetary loss due to the information contained within this book, either directly or indirectly.

Legal Notice:

This book is copyright protected. It is only for personal use. You cannot amend, distribute, sell, use, quote or paraphrase any part, or the content within this book, without the consent of the author or publisher.

Disclaimer Notice:

Please note the information contained within this document is for educational and entertainment purposes only. All effort has been executed to present accurate, up to date, reliable, complete information. No warranties of any kind are declared or implied. Readers acknowledge that the author is not engaged in the rendering of legal, financial, medical or professional advice. The content within this book has been derived from various sources. Please consult a licensed professional before attempting any techniques outlined in this book.

By reading this document, the reader agrees that under no circumstances is the author responsible for any losses, direct or indirect, that are incurred as a result of the use of the information contained within this document, including, but not limited to, errors, omissions, or inaccuracies.

ISBN: 978-1-7638713-0-4 (paperback)

Published by Nexus Global Media

Email: contact@nexusglobal.media

ACKNOWLEDGMENTS

Life's greatest teachers often disguise themselves as our biggest challenges. As I reflect on the journey that led to this book, I am profoundly grateful for both the light and the shadow moments that shaped my path to self-love.

First, I want to acknowledge the experiences that brought me to my knees—the relationships that shattered my heart, the anxiety that gripped my days, the depression that clouded my vision, and every moment when I felt lost and disconnected from myself. These challenges weren't just obstacles; they were invitations to dig deeper, seek answers, and ultimately discover the transformative power of self-love. Without hitting rock bottom, I might never have embarked on this journey of discovery and healing.

To Dr. Espen Wold-Jensen, whose wisdom and guidance illuminated the path when I needed it most—thank you for showing me how to bridge the gap between science and soul, helping me understand that transformation isn't just about feeling better but about rewiring our very biology for lasting change.

To every teacher, mentor, and guide who shared their knowledge and insights—your work has been instrumental in my personal transformation and the creation of this book. I would like to give a special acknowledgment to Dr. Joe Dispenza, whose groundbreaking work in neuroplasticity provided a scientific foundation for understanding how we can rewire our brains for self-love.

To my mother, whose presence I feel guiding me even as she soars with her angel wings—your struggles and pain have been woven into the fabric of my own journey. Through you, I learned both the weight of generational trauma and the profound possibility of healing it. In facing my own darkness, I came to understand yours, and in healing myself, I heal us both. The work we do together now transcends the physical realm, transforming not just my path but yours, too. The cycles of pain that you endured have become the catalysts for transformation in our lives. This work of self-love is as much your journey as it is mine—a testament to the truth that while we inherit our wounds, we also inherit the strength to heal them.

To my children, who inspire me daily to embody the self-love I write about—your presence in my life has deepened my understanding of unconditional love and strengthened my commitment to breaking generational cycles. By choosing to heal myself, I also choose to show you a different way. Through my journey, you're learning that self-love isn't just something we talk about—it's something we live, breathe, and practice daily. My commitment to this path is as much for you as it is for me and your grandmother, ensuring that what you inherit is not the weight of past wounds but the wisdom of self-love and the strength to maintain it.

To every client who has trusted me with their own journey—your courage in facing your own challenges has inspired and taught me more than you'll ever know.

And perhaps most importantly, I want to acknowledge my younger self—the one who felt unworthy, who struggled to speak her truth, who put everyone else first. Your pain led us here. Your resilience kept us going. Your willingness to feel it all made this transformation possible.

To every reader who picks up this book, may you find within these pages what my challenges taught me—that every struggle carries within it the seeds of transformation and that the path to self-love, while not always easy, is always worth taking.

With deep gratitude,

Fiona

TABLE OF CONTENTS

Introduction 9

Chapter 1: The Hidden Pattrens—Unmasking the Signs of Low Self-Love...... 23

Chapter 2: Awareness—The First Step to Self-Love 37

Chapter 3: Rewiring Your Brain for Self-Love 53

Chapter 4: Releasing Emotional Blocks to Self-Love.. 73

Chapter 5: Embracing Your Inner Child.. 89

Chapter 6: Cultivating Gratitude as a Gateway to Self-Love........................... 101

Chapter 7: Setting Boundaries as a Act of Self-Love 115

Chapter 8: Physical Self-Love—The Gentle Awakening 133

Chapter 9: Making Self-Love Sustainable—From Practice to Way of Being.. 155

Conclusion: From Practice to Transformation 177

Resources.. 181

About the Author 186

Until you love yourself,
you will never know
WHO YOU REALLY ARE,
and you won't know
WHAT YOU'RE REALLY
CAPABLE OF.

LOUISE HAY

INTRODUCTION

Self-love.

It's a term that appears everywhere, from the self-help books flooding our bookstores to the #selflove hashtags on social media.

But let's be honest for a moment: How often do you stop to think about what self-love actually means? Is it treating yourself to that "thing" you've been wanting, a spa day, or an evening binging on your favorite show? Or is it something more complex that could fundamentally change how you show up in every aspect of your life—from your relationships and career to the quiet moments when you're alone with your thoughts?

I discovered the true depth of this question after a painful pattern of relationships that seemed to repeat themselves. Several years ago, after yet another unhealthy partnership, I fell into the pit of depression and acute anxiety. Like many, I first sought help through traditional pathways—seeing psychologists and psychiatrists and trying various pharmaceuticals. But despite following all the conventional advice, I felt stuck in a cycle of despair, unable to find my way back to myself.

During this dark period, I started reflecting deeply on a startling truth: The unhealthy patterns weren't just present in my romantic relationships, they were everywhere in my life—and I was the common denominator.

I was the person who never said no, who smiled and agreed while screaming inside, who put everyone else's comfort before my truth. I was also the person who consistently chose partners who dominated the relationship while I remained silent, unable to maintain boundaries and eventually realizing that my needs were constantly being pushed aside in favor of keeping the peace. This pattern of self-abandonment is one that many of us know too well, and it can show up in many different ways.

When this awareness finally landed, it led me to ask crucial questions: Why was I consistently choosing and allowing these toxic relationships? Why did the thought of expressing my needs fill me with such intense anxiety? Why was I being a people-pleaser? And, most importantly, where was my self-love in all of this?

I'll be honest, these questions were confronting, but through this raw self-reflection I began to understand that these patterns weren't random. At their core was an acute inability to set healthy boundaries and a deep-seated fear of conflict that kept me mute when I should have spoken up. Like many who struggle with self-love, I had mastered the art of putting everyone else first. This tendency meant sacrificing my well-being to keep others happy. The painful truth was that I had learned to love others at the expense of loving myself.

These weren't just bad habits—they were deeply rooted programs that I finally understood had been installed during my childhood, shaped by growing up in an environment where love felt conditional and speaking up felt dangerous.

This startling realization led me down an unexpected path, where ancient wisdom met modern neuroscience. A chance conversation led me to discover neuroscience experts who were connecting their findings to ancient wisdom, and I finally began to understand what had eluded me—why I kept finding myself in relationships and situations that caused me pain. My patterns of self-abandonment weren't just psychological; they were wired into my brain through neural pathways. But here's the incredible thing these experts taught me: These neural pathways aren't permanent. They can be rewired. This understanding became my beacon of hope, showing me that, no matter how deeply ingrained my patterns of self-doubt, people-pleasing, and silence were, they could be transformed.

NEUROSCIENCE APPROACH AND SELF-LOVE FOUNDATION

Before we begin, I want to be clear: I'm not a scientist or a researcher. I'm someone who went searching for answers and found them through the work of brilliant experts who have dedicated their lives to understanding how we can heal and grow. What I share in these pages comes from my personal journey of discovery, supported by the wisdom and research of those who helped light my path.

My turning point came unexpectedly. After years of struggling with anxiety and depression, searching through countless self-help podcasts without finding sustainable solutions, a conversation with a friend led me to revisit the documentary *What the Bleep*

Do We Know? In this documentary, Dr. Joe Dispenza—a leading neuroscientist, best-selling author, and pioneer in the fields of quantum physics and neuroplasticity—explained how quantum physics connected science and spirituality and, more importantly, how we could actively participate in rewiring our neural pathways. This resonated with me deeply, as if someone had turned the lights on in my mind. As someone who had never considered herself "scientific," this was revolutionary. Here was an approach that made sense—one that validated ancient wisdom through the lens of modern science.

This discovery changed everything. I began to understand that what spiritual traditions had advocated for millennia wasn't just mystical thinking—it had concrete, scientific backing. The practices these traditions recommended—meditation, journaling, conscious awareness—were being validated by scientific research. This wasn't just about feeling better—it was about literally rewiring my brain for lasting change.

Through studying Dispenza's work, I discovered how our brains constantly evolve, creating new neural pathways based on our experiences, thoughts, and actions. This understanding of neuroplasticity offered a significant hope: No matter how deeply ingrained my patterns of self-doubt and people-pleasing were, they could be transformed through intentional practices like affirmations, visualization, meditation, and mindfulness. These weren't just theories—they were scientifically proven tools that had helped countless people, including me, rebuild their relationship with themselves.

Dispenza's research shows that intentional practices such as affirmations, visualization, meditation, and mindfulness can reshape our neural networks. These aren't just theories—they're scientifically proven tools that have helped countless people, including me, rebuild their relationship with themselves.

SELF-LOVE: THE MOST IMPORTANT LOVE OF ALL

> *"Love is a way of being; love is actually present everywhere, it's present only needs to be realized… love is the energy that silently transfigures every situation."*
> **–David R. Hawkins**

Let's be clear: Self-love isn't selfish or about being vain.

Many of us, especially those who identify as caregivers or nurturers, have been conditioned to view self-care as selfish. We've learned to dim our light, silence our voice, and shrink ourselves to make others comfortable. But this couldn't be further from the truth. True self-love is about treating yourself with the same care, kindness, and compassion you so readily extend to others. It's about honoring your needs, setting (and maintaining) healthy boundaries, and recognizing your inherent worth—whether that's speaking up in a meeting, saying no to extra commitments, or choosing relationships that celebrate your value.

In today's fast-paced world, where we're constantly bombarded with messages about not being enough, understanding true self-love becomes even more crucial. Here's another way to think about it: Imagine your love for others as a light shining out into the world. For that light to burn brightly, it must first have a steady source of fuel. Self-love is that fuel. It sustains you so you can show up fully for the people you care about. Without the foundation of self-love, even the strongest relationships can falter.

Another consequence of lacking self-love is an endless search for external validation. I know this pattern intimately—I spent years looking for others to fill the emptiness I felt inside. It showed up everywhere: seeking approval from my partners, working overtime to prove my worth to my employers, saying yes to every request to be seen as "helpful," and constantly adapting my behavior to please others. However, external validation is like trying to fill a leaking bucket: No matter how much others pour in, it will never be enough until we repair the foundation of self-love within ourselves.

It's important to note that our relationship with self-love is deeply influenced by our background, family dynamics, and life experiences. What self-love looks like for you might be different from how it appears to others, and that's perfectly okay.

YOUR JOURNEY STARTS HERE: WHO THIS BOOK IS FOR AND WHAT TO EXPECT

The journey to loving yourself looks different for everyone. While my wake-up call came through unhealthy romantic relationships, yours might come from

- ➢ burnout from constantly putting everyone else's needs first.
- ➢ health issues that force you to finally prioritize your well-being.

- a career situation where you've lost your voice.
- family dynamics that have taught you to minimize your needs.
- the exhaustion of maintaining a perfect image for others.

The entry point matters less than recognizing the pattern. For some, the pattern manifests as chronic anxiety, for others as an inability to say no, and for many, like me, as a series of relationships that reflect our lack of self-worth. What matters is that you're here now, ready to understand and transform these patterns.

You might be thinking, *I've tried self-help books before, and nothing changed*. I understand this feeling. Traditional self-help tools like affirmations and journaling are powerful practices and an essential part of the transformation process. However, they're most effective when combined with a deeper understanding of how our brain creates and maintains patterns.

That's why this book offers a comprehensive approach. By combining the science of neuroplasticity with proven self-development practices, you'll learn what to do as well as why and how these practices work to create lasting change. Understanding the neuroscience behind how your brain forms patterns allows you to use tools like affirmations and journaling more effectively as part of a complete system for rewiring your neural pathways.

And if you're thinking, *I barely have time for another self-improvement project*, I get that too. The strategies in this book are designed to work with your life, not add to your overwhelm. They're practical, science-backed approaches that can be integrated into your existing routine while supporting real, lasting transformation.

This book is both a guide and a tool kit. Inside, you'll find

- **scientific insights:** learn how neuroscience and other frameworks shape your ability to love yourself.
- **practical strategies:** discover tools you can use daily to help you cultivate self-love, from journaling prompts to guided meditations.
- **spiritual wisdom:** tap into teachings from leading experts in the field that connect you with your most authentic, loving self.
- **actionable change:** turn insights into habits that transform your brain, your energy, and your life.

By the end of this book, you'll have

- the neuroscience knowledge to understand how your brain can change.
- practical tools to set and maintain healthy boundaries without guilt.
- strategies to quiet your inner critic and build lasting self-worth.
- techniques to manage anxiety and stress through self-love practices.
- methods to balance caring for others while honoring your own needs.
- a personalized road map for your ongoing self-love journey.

These aren't just theoretical concepts I've read about. Throughout my journey, I've personally tested and practiced each strategy in this book. I've applied them in romantic relationships, in professional settings where I needed to find my voice, in family dynamics where I needed to set boundaries, and in daily moments where I needed to choose myself instead of defaulting to people-pleasing.

Through this process, I've learned something profound: From understanding how your brain forms self-love habits to releasing the emotional blocks that keep love from flowing freely, this journey is about building a lasting foundation of self-worth. These practices have transformed not just my relationships but every aspect of how I show up in the world.

As someone who's walked this path not as an expert but as a seeker, I understand the challenges of this journey. I've spent countless hours studying the work of leading researchers and practitioners, translating their scientific insights into practical, accessible tools. While I'm not a scientist, I am a dedicated student and practitioner of these transformative approaches. My role in writing this book is to share what I've learned in a way that makes sense for real life—your life.

It's important to note that this process isn't about quick fixes or surface-level solutions. It's not about suddenly becoming a different person or never struggling again. Instead, it's about creating sustainable change by understanding and applying the science of how our brains work while honoring our unique journey.

ARE YOU READY?

Self-love isn't a destination; it's a journey. It's about showing up for yourself, even when it feels hard. It's about making choices that align with your worth and reminding yourself that you're enough, just as you are. And because this is a journey, every step you take is a step toward the life and self-love you deserve.

So, let's take that first step together.

NOTE TO THE READER

> *"In science, we call it energy. In religion, we call it spirit. On the street, we call it vibes. All I am saying is… trust it."*
> —Bruce H. Lipton, PhD

Before we begin, I want to acknowledge something important. Whether you come from a spiritual background or a purely scientific one, you might find yourself skeptical about some aspects of this book. That's not just okay—it's welcome.

My own journey began with an intuitive draw toward spiritual concepts, yet I lacked the scientific understanding to explain why certain practices worked. When I discovered how neuroscience validated these ancient practices, it was like finding the missing piece of a puzzle. This bridge between ancient wisdom and modern science transformed my understanding of self-love and healing.

If you're coming from a scientific perspective, you might have initially dismissed practices like meditation, journaling, or conscious breathing as lacking empirical evidence. I get it. However, the field of neuroscience has not only made remarkable strides in understanding how our brains work but has also provided substantial evidence supporting these practices. Studies have shown that regular meditation can increase gray matter density in the regions of the brain associated with emotional regulation, learning, and memory (Hölzel et al., 2011). Research has also documented how mindfulness practices can reduce stress hormones, strengthen immune function, and even influence gene expression (Tang et al., 2020). These findings don't just validate scientific principles—they demonstrate what spiritual traditions have known for millennia.

Some might question whether neuroscience is a pseudoscience. Yet what was once considered fringe research has evolved into a robust field of study, with neuroplasticity now established as a fundamental principle of brain science. The evidence is compelling: Our brains can form new neural connections, adapt to experiences, and, in certain regions, even generate new neurons well into adulthood. Articles like "Adult Neuroplasticity: More Than 40 Years of Research" in the journal *Neural Plasticity* (Fuchs & Flügge, 2014) document how specific practices create measurable changes in brain structure and function. If you're curious to learn more, I encourage you to explore the scientific references listed at the end of this book.

On the other hand, if you're more spiritually oriented, you might question whether bringing science into these practices somehow diminishes their sacred nature. What I've discovered is that understanding the neuroscience behind these ancient practices actually deepens their significance. Science isn't replacing spiritual wisdom; it's providing another lens through which to understand these transformative practices. When we see how meditation reshapes our brain or how conscious breathing affects our nervous system, it reinforces what these timeless practices have always offered: a powerful transformation of our whole being.

This book represents a union of these two worlds. Through my personal journey and extensive research, I've found that science and spirituality aren't opposing forces—they're different languages describing the same transformative truths. The practices I share aren't just theoretically sound; they're practical tools that have helped countless people, including myself, create lasting change.

I invite you to approach this book with curiosity, regardless of your starting point. You don't need to accept everything at face value. Try the practices, explore the science, and discover what works for you. This is your journey, and I'm simply here to share what I've learned along the way.

The masks we wear to seem **'strong'**
often hide the wounds
we fear to face.
TRUE STRENGTH BEGINS
when we stop performing
for the world

and start showing up
for ourselves.

FIONA SOUTTER

CHAPTER 1

The Hidden Patterns—Unmasking the Signs of Low Self-Love

WHEN "HAVING IT ALL TOGETHER" IS A MASK

For most of my life, I didn't realize I lacked self-love. From the outside—and even to myself—everything appeared fine. Better than fine, actually. I was the person who was always there for others, achieving my goals and maintaining a smile no matter what life threw my way. These traits earned me constant praise. "You're so strong," people would say. I took pride in my ability to persevere through any challenge.

But our bodies and spirits have a way of demanding attention when we've ignored them for too long. For me, that moment came when I fell into a pit of depression and anxiety that I couldn't escape. This wasn't part of my plan. I was frustrated and confused—why couldn't I maintain my usual resilience? Why couldn't I simply put on my practiced smile and keep going? For the first time in my life, my well-honed adaptive behaviors weren't working.

What I didn't realize then was that this breakdown was actually a breakthrough. This pit wasn't just about the present situation—it was as if every suppressed emotion, every buried feeling, every ignored signal from my past had suddenly surfaced at once. The weight wasn't just from my current pain; it was the accumulation of all the pain I'd ever felt but never truly processed. It was the realization that what had initiated the spiral was only one instance in a life full of recurring patterns. This was what made it different. This was why I couldn't simply rise and continue as I had so many times before.

THE PATTERNS WE ALL SHARE

Perhaps you recognize these thoughts:

- Why does this always happen to me?
- I keep attracting the same type of toxic relationships.
- Nothing ever works out in my favor.
- Bad things just seem to find me.

These recurring patterns are often the first hints that something deeper is at play. While we might attribute these experiences to bad luck or external circumstances, there's a more profound truth: These patterns are precise reflections of our internal state.

Many people caught in these cycles aren't aware that their perceived "unfortunate circumstances" are manifestations of hidden patterns of low self-love. They might notice the surface-level similarities—another demanding boss, another controlling partner, another financial setback—but miss the underlying thread that connects them all.

UNDERSTANDING THE SCIENCE OF PATTERNS

> *"If you want to find the secrets of the universe, think in terms of energy, frequency, and vibration."*
> **—Nikola Tesla**

The patterns that show up in our lives aren't just psychological—they operate at a deeper, quantum level. Modern physics has shown us that everything in the universe, including our thoughts and emotions, exists as energy vibrating at different frequencies. Tesla's understanding of energy and frequency, which was revolutionary for his time, has been validated and expanded by modern quantum physics, revealing how our internal state directly influences our external reality.

Building on this quantum understanding, Dispenza's groundbreaking work challenges the traditional Newtonian model of cause and effect (Howes, 2024). His research introduces us to a quantum understanding where our thoughts and emotions create electromagnetic fields that influence our reality.

This quantum perspective reveals how our recurring patterns become encoded in our neural pathways. Dispenza teaches that by repeatedly rehearsing specific thoughts and emotional responses, we condition our bodies to automatically respond to external stimuli in predictable ways (Shetty, 2023). These conditioned responses create recurring patterns in our lives. However, through understanding the intersection of neuroscience, quantum physics, and consciousness, we can break free from these automated patterns. By harnessing the power of our mind and consciously choosing new emotional states, we can literally reprogram our neural pathways and transform our reality.

With this understanding, we can begin the process of transformation. When we consciously choose elevated emotional states like love, joy, and gratitude, we create new neural pathways that override our old conditioning. These elevated states create changes that affect our entire being, from our cellular structure to our daily experiences.

Through his work as an international researcher and educator, Dispenza has documented numerous instances where individuals have created profound changes in their lives by understanding and applying these principles (Howes, 2018). The key lies in recognizing how our thought patterns become wired into our brain and learning how to create new, more beneficial patterns.

While Dispenza's work laid the foundation for my understanding of how thought patterns create our reality, it was a personal mentorship with Dr. Espen Wold-Jensen, renowned for his expertise in neuroscience and quantum physics, that accelerated my transformation, especially in overcoming deep, lingering emotional blocks. Dr. Espen illustrated how our vibrational frequency affects transformation with a simple yet significant diagram showing two parallel lines, one representing our current vibrational state and the other representing what we desire to attract or achieve (see Figure 1.1). These lines, being parallel, never meet. This visual finally helped me understand why nothing was truly shifting despite my desire for better relationships, improved health, and different life circumstances—the patterns of my internal frequency simply couldn't attract or sustain what I desired.

Figure 1.1: The frequency gap between what you want and where you are

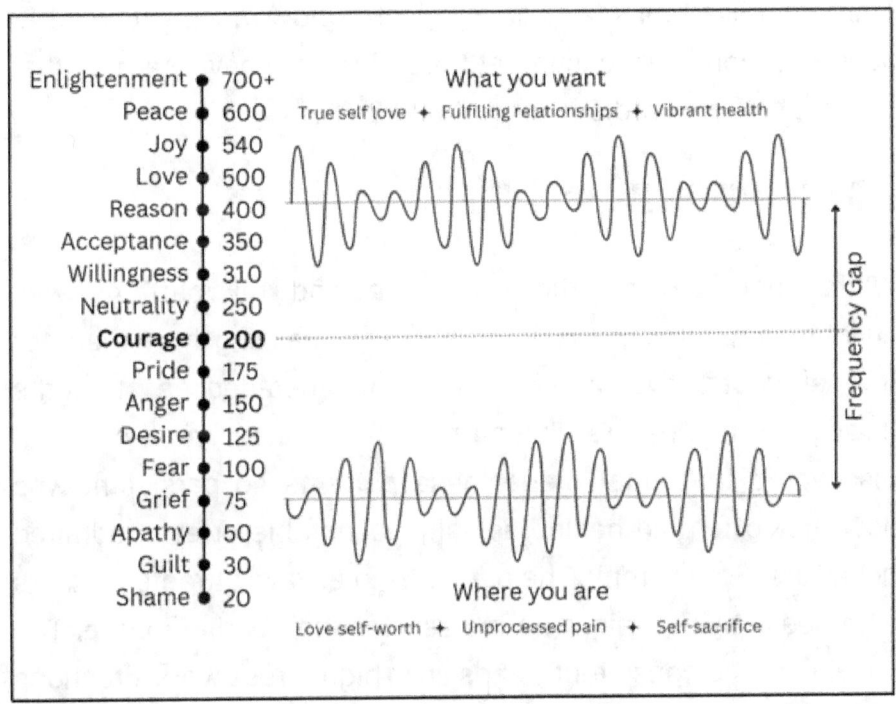

Through my work with Dr. Espen, I began to understand the deeper implications of this frequency alignment. He showed me how our core energy centers hold onto negative emotions and past traumas, creating energetic blockages that keep us stuck in old patterns (Wold-Jensen, 2019). Drawing from his expertise in neuroscience and quantum physics, he guided me through the practical application of these principles, helping me see how my own stored emotions were affecting my frequency.

As we'll explore in the next chapter, this understanding of how emotions operate at different frequencies became the key to finally shifting my patterns and creating lasting change. What began as an intellectual understanding through Dispenza's work transformed into lived experience through Dr. Espen's guidance. Think of it like tuning a radio—no matter how much you want to hear a station broadcasting at 98.5 FM, you'll never receive that broadcast if your radio is tuned to 92.3 FM.

WHEN STRENGTHS MASK STRUGGLES

While understanding the science behind my patterns was enlightening, what truly shook me was realizing how many of my self-destructive behaviors had masqueraded as strengths throughout my life. As I began to examine them more closely, I discovered that what made these patterns so persistent was how they often hid behind socially acceptable or even admired behaviors. For example, the person who always attracts partners who need "fixing" might be praised for their caring nature. The individual who consistently ends up in high-stress work situations might be admired for their work ethic. The friend who's always solving everyone else's problems while neglecting their own might be celebrated for their selflessness.

In my case, the ability to suppress my emotions and endure any situation, coupled with a lack of boundaries rooted in low self-worth, meant I consistently attracted circumstances and relationships that both demanded and exploited these traits. This left me extremely vulnerable to toxic dynamics. In the business world, it was praised as "resilience." In friendships, it made me the "strong one" whom everyone could rely on, when in reality, I had nothing left in the emotional tank for myself.

But operating at this frequency meant I couldn't attract or sustain relationships where vulnerability and authentic emotional expression were welcomed. The migraines that plagued me for years were my body's way of communicating that this frequency wasn't sustainable. When all my carefully constructed adaptations failed at once, I finally had to face the truth: What I labeled as strength was actually fear, what I called resilience was really avoidance, and what looked like having it all together was masking a deeply rooted disconnection from myself.

THE DAWN OF TRANSFORMATION

Understanding these hidden patterns and the science behind them offers more than just an explanation; it offers hope. Everything changes when we recognize that our struggles aren't character flaws but energetic patterns that can be transformed. The very awareness of how we're operating at a particular frequency opens the door to the possibility of change.

This shift doesn't happen through forcing change or adding more items to our self-improvement checklist. It begins with awareness— the capacity to observe these patterns without judgment and see them for what they are. Think of this understanding as switching on a light in a dark room. Suddenly, the patterns you couldn't see before become visible, and the behaviors you thought were supporting you reveal themselves as adaptations that no longer serve their purpose. This illumination, while sometimes uncomfortable, is the beginning of true transformation.

In the following chapters, we'll explore exactly how to develop this awareness and use it as a foundation for lasting transformation. We'll learn practical tools for shifting our frequency and creating sustainable change. But it all begins here—with the recognition that what we've been experiencing isn't random and that our patterns, no matter how deeply ingrained, can be transformed.

☆ A PRACTICAL EXERCISE: PATTERN RECOGNITION THROUGH ENERGY AWARENESS

One of the most powerful ways to begin transforming your patterns is to first become aware of how they manifest in your daily life, particularly through your energy states and recurring situations. This energy pattern mapping exercise is designed to help you identify and understand your unique patterns from an energetic perspective.

CREATE THE SPACE (15-20 MINUTES)

Find a quiet place where you can reflect without interruption. You'll need a journal or notebook and something to write with. This exercise works best when you can fully focus on sensing your energy and observing your patterns.

STEP 1: ENERGY BASELINE ASSESSMENT

Before diving into pattern recognition, take a moment to tune into your current energy state:

1. Take three deep breaths, placing one hand on your heart.
2. Notice how your body feels in this moment.

3. Observe your current emotional state.
4. Rate your energy level on a scale of 1–10.

STEP 2: PATTERN IDENTIFICATION

Create three columns in your journal:

- **Column 1: Recurring situations:** List situations that keep showing up in your life. Examples might include:
    - relationships that follow similar dynamics
    - work scenarios that feel familiar
    - financial patterns that repeat
    - health issues that cycle back

- **Column 2: Energy response:** For each situation, note:
    - How does your energy shift when you're in this situation?
    - What physical sensations arise?
    - What emotions surface?
    - Where in your body do you feel the response?

- **Column 3: Adaptive behaviors:** Observe how you typically respond:
    - What coping mechanisms do you use?
    - What "strength" do you demonstrate?
    - How do others perceive your response?
    - What praise or feedback do you receive?

Here's an example from my journal:

Recurring situation	Energy response	Adaptive behavior
Taking on additional work projects despite feeling overwhelmed	Chest tightening, shallow breathing, a mixture of anxiety and pride	Push through, maintain a composed exterior, receive praise for reliability

STEP 3: FREQUENCY CONNECTION

Now, look at your patterns through the lens of frequency:

- What energy state are you typically operating from in these situations?
- How might this frequency be attracting or maintaining these patterns?
- What would a higher-frequency response look like?

STEP 4: PATTERN INTEGRATION

Review what you've written and answer these questions:

- What common themes do you notice across different situations?
- How do your adaptive behaviors maintain these patterns?
- What praise or external validation keeps these patterns in place?
- How do these patterns protect you? What do they cost you?

ADVANCED PRACTICE: DAILY ENERGY TRACKING

For deeper insight, maintain an energy pattern log for one week:

- Note the situations that trigger energy shifts.
- Record your immediate adaptive response.
- Document any praise or validation you receive.
- Observe how the pattern maintains itself.

The goal isn't to judge or change these patterns yet but instead to simply to observe them with curiosity and compassion. Like a scientist studying a phenomenon, you're gathering data about how your energy system operates.

📌 KEY TAKEAWAYS:
RECOGNIZING HIDDEN PATTERNS OF LOW SELF-WORTH

- **External success ≠ internal worth:** What appears as "having it all together" can mask deep patterns of low self-love. Success, achievement, and being "the strong one" may be adaptive behaviors rather than true self-love.
- **Patterns are energy in motion:** Our recurring life situations aren't random—they're precise reflections of our energetic frequency. Understanding this connection is crucial for lasting transformation.
- **The body speaks the truth:** Physical symptoms, emotional breakdowns, and recurring health issues often signal that our adaptive patterns are no longer serving us. These aren't failures but opportunities for awareness.
- **Praise can perpetuate unhealthy patterns:** External validation for our adaptive behaviors (being "strong," "resilient," and "selfless") can keep us stuck in patterns of low self-love by reinforcing unhealthy coping mechanisms.
- **Quantum understanding brings hope:** Recognizing that our patterns operate at a quantum level means we can change them by shifting our frequency rather than through force or willpower alone.
- **Awareness precedes transformation:** Simply observing our patterns without judgment creates the space needed for genuine change. This awareness is the foundation for all lasting transformation.
- **Strength can mask struggle:** What we label as personal strengths may actually be sophisticated coping mechanisms developed in response to low self-love. True strength includes acknowledging our struggles.

> **Change begins with pattern recognition:** Understanding our unique patterns of low self-love—how they manifest, what maintains them, and how they protect us—is the first step toward transforming them.

Remember: Your patterns developed for valid reasons and served important purposes in your past. Approaching them with curiosity rather than criticism creates the conditions for genuine transformation.

There is nothing
more important
to true growth
than realizing you are
NOT
the voice of your mind

You are the one who
hears it.

MICHAEL SINGER

CHAPTER 2:

Awareness—The First Step to Self-Love

A transformative shift in my relationship with myself began at my desk one afternoon. As I prepared for an important client presentation, the familiar cascade of self-doubt emerged: *Are you really qualified to do this? What if they see through you?* But this time, something changed. Instead of being swept away by these thoughts, I found myself watching them unfold as if observing a scene from outside myself. This wasn't just another moment of self-reflection—it was my first glimpse of a fundamental truth. These thoughts weren't me. They were merely programs running in my mind.

Spiritual teacher and number-one *New York Times* bestselling author Michael Singer captures this distinction perfectly when he explains: "The very fact that you can see the disturbance means that you are not it. The process of seeing something requires a subject–object relationship. The subject is called the witness because it is the one who sees what's happening."

In that moment of recognition, I realized there was a distinct separation between my true self (the one doing the observing) and these automated thoughts that had been running unchecked for so long. It was like suddenly becoming aware of background music that had been playing my entire life—music I had mistaken for my own voice.

Like many, I had mastered the art of projecting confidence while these inner dialogues shaped my reality beneath the surface. But that initial moment of observing—of *seeing* rather than *being* these thoughts—changed everything. I wasn't the program; I was the one watching it run.

> "When you become aware that you are aware, that is awareness right there."
> —Dr. Espen Wold-Jensen

This recognition is essential when it comes to self-love. Most people struggling with self-esteem don't realize just how many negative thoughts they harbor about themselves. They feel the weight of these thoughts in their daily lives—in how they view themselves, their appearance, and their choices. But what if the core issue isn't who you are but rather these automated programs running in your mind?

As you'll recall from the Introduction, my journey into understanding thought patterns began unexpectedly, with a segment by Dr. Joe Dispenza in the documentary *What the Bleep Do We Know?* about the connection between quantum physics and consciousness. Through studying Dispenza's work, I discovered that our thoughts actively shape our reality. We generate tens of thousands of thoughts daily, most of which are unconsciously negative and self-critical. These automatic programs operate silently, coloring how we see ourselves without our awareness. Until we step back and observe them, we remain caught in their influence, mistaking these programs for our true identity.

The gift in this discovery? That moment of becoming the observer—when we recognize these thoughts as programs rather than truth—marks the beginning of our transformation.

THE POWER OF OBSERVING

After that initial recognition between my observing self and my thinking mind, I began detecting these thought programs everywhere in my life, far beyond my business concerns. It felt like I was carrying a backpack filled with automated responses and beliefs—patterns programmed by past experiences, cultural conditioning, and years of unconscious habits.

Picture carrying a backpack loaded with rocks. Each rock represents a programmed thought pattern or belief: *I'm not good enough, I need to prove my worth, I can't trust my judgment*. You've carried this weight for so long that it feels normal. The discomfort has become such a constant companion that you've forgotten what life felt like before these programs.

Now, imagine stepping back and seeing this backpack for the first time. In that moment of recognition, you realize something vital: This backpack isn't you. It's something you've learned to carry. And, just as you've learned to carry it, you can learn to set it down.

Singer beautifully illuminates this concept, describing awareness as light entering a dark room. When you detect these automated programs running—those negative thoughts and feelings—you create space between you (the observer) and the programs themselves. This space becomes the catalyst for change.

When you begin observing your thoughts rather than being them, patterns emerge in these mental programs. Like a background application constantly running on your computer, this program offers a commentary on everything you do.

DETECTING THE INNER CRITIC

When you begin observing your thoughts rather than being them, patterns emerge in these mental programs. One of the most persistent is the inner critic. Like a background application constantly running on your computer, this program offers a commentary on everything you do.

When I first stepped into the observer role, I was startled by how consistently this critical program operated. While I was building my business, it would interject with thoughts like, *You're not experienced enough for this client*, or *Other entrepreneurs are so much further ahead than you.* What struck me was that I could now watch these thoughts arise without becoming entangled in them. From this observing perspective, I recognized them as just another program, not the truth of who I am.

Dr. Justine Grosso, a psychologist specializing in emotional resilience, explains that this inner critic often develops as a protective mechanism. Its original purpose was to shield us from failure, rejection, or disappointment by keeping us "in line." However, like outdated software that no longer serves its purpose, this critical voice can become a significant obstacle to self-love.

Singer adds another crucial insight: The key isn't to fight these critical thought patterns, as attempting to battle them only creates more internal conflict. Instead, from your position as observer, simply watch them. Notice how they arise, what triggers them, and how they attempt to influence your behavior. This neutral observation naturally weakens their power as you no longer identify them as "you."

While observing the inner critic marks an important first step, it often leads us to discover an even deeper layer of programming: our self-limiting beliefs. These beliefs operate in more subtle ways, influencing every aspect of our lives.

THE HIDDEN INFLUENCE OF SELF-LIMITING BELIEFS

Self-limiting beliefs exist on two levels: those we're conscious of—the daily negative self-talk we're all too familiar with—and those buried so deep we don't even know they're there. While I was well aware of my inner critic's regular commentary about me not being "good enough" as a business owner, I had no idea about the deeper beliefs silently shaping my reality until one ordinary afternoon.

I had posted my goals throughout my home and office, including a specific financial target for my business, and had followed the practice of writing them in the present tense. As I walked past these familiar words, my inner dialogue whispered something that stopped me in my tracks: *Yeah, but you probably won't achieve that.*

Unlike the obvious self-doubt I'd observed earlier in my journey, this thought was so subtle and quick that I almost missed it. It wasn't part of my usual conscious negative self-talk—it was something deeper and more ingrained. I paused to examine this whisper, asking myself, *Why did I think that?* The answer that surfaced shocked me: *Because success is something other people achieve.*

This quiet moment unveiled a belief system I had no idea I was carrying. Looking back, I could trace its origins to my childhood. While my family was solidly middle-class, I grew up in an affluent neighborhood where most families lived in a different financial stratosphere, with sprawling houses, exotic holidays, and latest-model luxury cars. Through my child's eyes, the world seemed starkly divided between the "haves" and "have-nots"—and, despite our comfortable life, I unconsciously placed my family in the latter category.

My young mind, unable to grasp the nuances of socioeconomic status, had created a simplistic story: Money and success belonged to other people, not to me or my family. This belief had silently shaped my relationship with money for decades, yet I'd never consciously acknowledged its existence.

Here I was, consciously setting goals and taking actions toward them, while this hidden program ran beneath the surface, quietly undermining everything I was trying to achieve. Unlike the self-doubt I could easily identify—the "not good enough" thoughts that would surface during presentations or business meetings—this belief had operated completely outside my awareness.

As I paid closer attention, I noticed how this core belief manifested in subtle self-sabotage: procrastinating on business-growing activities, hesitating to take necessary steps toward financial goals, and finding reasons why "now wasn't the right time" to implement strategies I knew could increase my income. These weren't conscious choices but automatic responses driven by a belief system I didn't even know I had.

These deeper self-limiting beliefs operate like stealth software, running so quietly in the background that we rarely notice their influence. They're different from our obvious negative self-talk because they don't announce themselves; they simply direct our actions through almost imperceptible thoughts that flash through our minds countless times each day. While we might easily recognize and challenge our conscious self-doubt, detecting and transforming these hidden beliefs require a different level of awareness.

THE LANGUAGE OF EMOTIONS

During my journey of self-discovery, I encountered the work of Dr. David Hawkins, a renowned psychiatrist and consciousness researcher. Through over 250,000 calibrations during his 20-year study, he demonstrated how our emotional states could be mapped as measurable energetic frequencies. His book *Power vs Force* revolutionized my understanding of these emotional patterns.

Through my journaling practice, I had already begun tracking the origins of my self-limiting beliefs and sabotaging behaviors. But Hawkins' scale of consciousness helped me understand something deeper—the emotional frequencies that were driving these patterns. I was shocked to discover that beneath my outward appearance of confidence, love, and joy, I had been operating primarily from suppressed anger, fear, guilt, and shame.

These hidden beliefs don't just influence our thoughts and actions—they create specific emotional resonances that Hawkins mapped on a logarithmic scale. He showed how we move from these lower vibrations through courage and acceptance, rising toward the higher frequencies where we find reason, love, joy, peace, and enlightenment. Understanding this scale helped explain why my life kept cycling back to similar situations instead of maintaining the positive changes I desired: My underlying emotional frequency was acting like a magnet, pulling me back to experiences that matched those suppressed lower vibrations.

Think of it this way: Just as your phone picks up different radio frequencies, your body resonates with different emotional frequencies. I experienced this firsthand when I wanted to discuss a minor issue with my partner. Before I opened my mouth to speak, I noticed an uncomfortable sensation in my stomach—a tightness I might have previously ignored or pushed aside. But instead of dismissing it, I used my observer perspective to tune into this frequency, like adjusting a radio dial to clarify a fuzzy signal.

As I stayed with the sensation, I recognized it as fear. Rather than pushing forward with the conversation, I watched how this fear was connected to whispered thoughts running in the background of my mind: *They'll leave if you make this an issue* and *You're too needy*. From my observer perspective, I could see these thoughts stemming from a deeper limiting belief: *I'm not lovable*. This awareness helped me understand that my hesitation wasn't really about the minor issue I wanted to discuss—it was my old programming running its familiar pattern.

Drawing on Hawkins' teaching that we can choose to operate from higher frequencies, I asked myself a simple but powerful question: *How would love respond?* This question created a shift. Instead of letting the fear program run its usual course, I acknowledged it and consciously chose to move into a space of love. The tight sensation in my stomach dissolved, and from this higher frequency, I could have a healthy conversation with my partner. I communicated my boundaries clearly while coming from a place of love rather than fear. Not only did the conversation not end in conflict, it also strengthened our relationship. More importantly, I practiced self-love by choosing not to let an old fear program dictate my actions.

As this experience showed me, the practice of tuning into our emotional frequencies gives us vital information about which programs are currently operating. When you feel emotions from the lower end of the scale (e.g. anger, fear, guilt, shame, grief), it signals that a limiting program is active. But here's where the power of being the observer becomes transformative: Instead of being these emotions or fighting against them, you can watch them arise, recognize them as signals, and consciously choose to operate from a higher frequency (e.g. acceptance, willingness, love, joy, peace).

For many of us, our initial response to these low-vibrational emotions is to suppress or ignore them. I did this for years—

pushing down feelings of inadequacy or fear, trying to "power through" them. But as Hawkins explains, suppressed emotions don't disappear; they become stored within us, creating energetic blocks that keep us stuck in these lower frequencies. The key is to recognize these emotions not as enemies to be fought or ignored but as messengers pointing us toward the programs that need our awareness. When we learn to observe these emotional frequencies and consciously choose to operate from higher ones like love, we begin to transform our entire experience of life.

☆ PRACTICAL EXERCISE: THE OBSERVER JOURNAL

One of the most powerful tools I've discovered for strengthening your observer perspective and uncovering both obvious and subtle programming is journaling. This practice helps you track your present-moment experience while reflecting on situations where you notice these programs running. Its flexibility makes it invaluable—you can engage with it whenever and however it serves you best.

CREATE THE SPACE (5-10 MINUTES)

Find a quiet place where you won't be interrupted. This matters because you're practicing stepping into your observer role, which requires some initial stillness. While I often do this in the morning, there's no "right" time. Let your natural rhythm guide you.

STEP 1: SET YOUR OBSERVER INTENTION

Before writing, take several deep breaths and remind yourself that you are not your thoughts or emotions—you are the one watching them. This simple reminder helps maintain that crucial observer perspective.

STEP 2: CHOOSE YOUR FOCUS

You have two powerful options:

- **Present-moment observing**
  - What thought programs are currently running?
  - What emotions or physical sensations are present?
  - How are you responding to what's happening now?

- **Recent experience reflection**
  - Recall a situation that triggered strong thoughts or emotions.
  - What programs were running during that experience?
  - What patterns can you see more clearly now?

STEP 3: DOCUMENT YOUR OBSERVATIONS

Whether you're observing the present moment or reflecting on a recent experience, write from your observer perspective:

- **For thought patterns**
  - "I'm watching the thought that says..."
  - "This program seems to believe..."
  - "I notice myself thinking..."

- **For self-limiting beliefs**
  - "When I think about [goal/desire], I observe the thought that..."
  - "This limiting belief appears to protect me by..."
  - "I notice resistance arising when..."

- **For emotional patterns**
  - "I'm watching these emotions surface when..."
  - "My body responds to these thoughts by..."
  - "I notice the connection between..."

Here's an example from my journal reflecting on a relationship interaction:

Witnessing my reaction to last night with my partner. I see how, after expressing my feelings and doing something special for them, I was seeking their validation. I notice the program that activated when their response wasn't as emotional as I hoped—immediately jumping to thoughts of "they don't love me as much as I love them" and "I'm not enough." From this witness perspective, I can track the familiar pattern: express love → seek validation → don't receive it in the expected way → spiral into unworthiness. I see how this same program has run in previous relationships, always searching for external proof of being loveable.

And here's an example from a present-moment observation:

Sitting here now, I watch thoughts arising about the upcoming launch. I see the familiar worry program running and notice how my breathing changes in response. From this witness space, I can track how this unfolds without getting caught in it.

STEP 4: REFLECT FROM YOUR OBSERVER SPACE

After writing, review what you've documented. Look for

- recurring thought programs.
- patterns in your emotional responses.

- triggers that activate specific programs.
- the space between you (the observer) and these patterns.

Let your journaling practice flow naturally. Some days, you might explore a specific situation; other days, you might observe what's present in the moment. Both approaches strengthen your observer perspective and help you recognize the difference between you and the programs running in your mind.

Remember, there's no "right" way to do this. The key is maintaining your observer perspective, whether you're watching programs run in real time or reflecting on how they performed in past situations.

THE OBSERVER'S GIFT

The journey from being our thoughts to observing them marks a fundamental shift in our relationship with ourselves. As we strengthen our observer perspective, we begin to see these automated programs—whether they're obvious critics or subtle saboteurs—for what they truly are: learned patterns rather than fundamental truths about who we are. This awareness creates the space needed for genuine self-love to emerge, not by fighting or fixing these patterns but by simply witnessing them with compassion and understanding. When we can step back and watch these programs run without being caught in their stories, we discover a simple truth: Beneath all these automated thoughts and beliefs lies our true self—the peaceful, aware presence that has been watching all along.

📌 KEY TAKEAWAYS:
BECOMING THE OBSERVER THROUGH AWARENESS

- You are not your thoughts—you are the one watching them. This recognition forms the foundation of true awareness and is the first step toward self-love.

- Your mind runs automated programs of thoughts and beliefs, installed in your childhood and through past experiences. Observing these programs is essential for transformation.

- The inner critic represents just one of these programs, not your authentic voice. By watching it rather than identifying with it, you begin to loosen its grip.

- Self-limiting beliefs often operate subtly, requiring careful attention to detect their influence on your choices and actions.

- Emotions serve as indicators of which programs are running. Rather than suppressing them, watch them with curiosity and allow them to move through you.

- Journaling from an observer perspective helps you identify patterns in your thoughts, beliefs, and emotional responses.

- The practice of observing creates space between you and your programming, allowing natural transformation to occur.

Remember: Change begins with awareness. You don't need to fight or fix your thoughts and emotions; simply observing them creates the space for transformation.

Self-love is an ocean,
and your heart is a vessel.
Make it full,
and any excess will spill over
into the lives of the people
you hold dear.

But you must come **first.**

BEAU TAPLIN

CHAPTER 3:

Rewiring Your Brain for Self-Love

Once I learned to observe my thought patterns and emotional programs, I faced a crucial question: Now what? Although being aware of these patterns was enlightening, I still found myself trapped in cycles of self-doubt and people-pleasing behaviors. I could watch the programs running, but they were still running.

That's when my journey led me to something remarkable: the science of neuroplasticity. I discovered that I could not only observe these mental programs but also change them. Just as I had learned to step back and watch my thoughts, I could learn to create new ones consciously.

Consider this perspective-shifting insight from Dispenza's research: Your brain processes around 400 billion bits of information every single second, yet your conscious awareness only registers about 2,000 bits. This means the vast majority of your life—including your self-talk and emotional responses—operates on autopilot, cycling through the same routines and creating repetitive experiences in your brain and body.

Through a process called synaptic pruning, we can actively fade away old neural connections while forging new ones. The science behind this is fascinating: Your brain contains approximately 100 billion nerve cells, each capable of forming up to 10,000 connections with other cells. This vast neural network—containing more potential connections than there are stars in the known universe—gives us an extraordinary capacity for change. In his groundbreaking work, Dispenza explains: "Neuroplasticity is our brain's ability to change its synaptic wiring by learning information and recording experiences... and to maintain a modified state of being."

Think of your mind like a forest, where your most frequent thoughts create well-worn trails through the wilderness. The self-doubt route in my mental landscape was practically a highway—smooth, wide, and so familiar I could travel it blindfolded. Each time I encountered a challenge in my business or relationships, my thoughts would automatically race down this familiar route: *You're not qualified enough, You're going to mess this up,* or *You're not worthy of love.*

A fundamental principle in neuroscience states that "what fires together, wires together"—the more you think a thought, the stronger that neural connection becomes. Creating new neural pathways is like blazing a trail through uncharted territory. When I first started intentionally choosing different thoughts—like *I am capable, I trust my abilities,* or *I am worthy of love*—it felt like trying to cut through dense undergrowth. It was uncomfortable and unfamiliar, and it required conscious effort. I often felt like I was "faking it" or that it wasn't working. But this is perfectly normal when creating neural connections that never existed before—it takes time and persistence to forge these new paths, as illustrated in Figure 3.1.

Figure 3.1: Rewiring neural pathways: a progressive journey

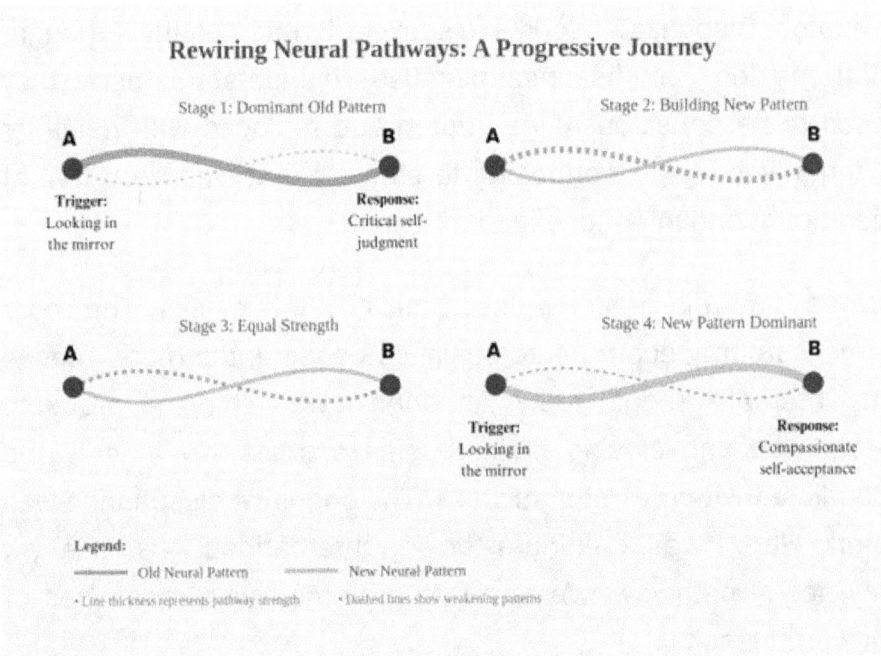

Through consistent practice, these new neural pathways begin to strengthen. What starts as pushing through thick mental undergrowth gradually transforms into a clear, accessible path to self-acceptance,

BREAKING THE CYCLE OF NEGATIVITY

Understanding that we can create new neural pathways is powerful, but the real transformation comes from actively breaking old thought patterns and establishing new ones. I learned this lesson the hard way in my ecommerce business. Even after understanding the science, I would still find myself spiraling into familiar patterns of self-doubt, especially during high-stakes moments like negotiating with international suppliers or securing business loans for expansion.

Through Dispenza's work, I discovered that negative thought patterns don't just disappear because we understand them—they need to be consciously interrupted and replaced with healthier alternatives. It's like choosing to step off that well-worn trail of self-doubt and intentionally forging a new path.

Here's how this process played out in my business. The more successful my ecommerce business became, the more intense my internal struggle grew. Although I was shipping products worldwide and closing major business deals, my inner world conflicted with my outer success. The cognitive dissonance was stark: Here I was, running a thriving international business, yet my neural pathways were still programmed with messages of unworthiness.

Using the observer perspective we explored in Chapter 2, I began to notice how these thought patterns would arise, particularly during crucial business moments. When negotiating a large inventory purchase or considering a significant business loan, I would catch thoughts like, *You don't belong in this league*, or *You're not a "real" business person*.

Instead of letting these thoughts run their usual course (which typically led to panic, physical stress symptoms, and crisis-mode decision-making), I learned to pause and engage in a three-step process:

1. **Identify the thought pattern:** I would name precisely what was happening: *Ah, there's my "imposter" program running again.* This simple act of identification, performed from the observer's perspective, already began to weaken the old neural pathway.

2. **Challenge the belief:** This is where I would question the thought pattern: *Is this actually true? Haven't I already successfully managed international suppliers? Haven't I already built a profitable business?* The evidence of my success directly contradicted these old belief patterns.

3. **Choose a new response:** This was the crucial moment—the point where I consciously stepped onto the new path. This meant replacing thoughts like *You don't belong here* with *My success is not an accident. I've earned my place at this table.*

I remember a pivotal moment when I successfully interrupted this pattern. I was about to secure a significant business loan for expansion—the kind of high-stakes decision that would typically trigger my unworthiness program and send me into panic mode. When those familiar thoughts and physical anxiety symptoms started, instead of spiraling into crisis, I paused, observed the pattern, and consciously chose a different response. I reminded myself that the success of my business wasn't a fluke but the result of my capabilities and hard work.

This experience taught me something crucial about rewiring our brains: Our thoughts aren't abstract concepts but actual neural pathways that influence our decisions and physical well-being.

Whenever we choose a new thought pattern, we're not just thinking differently; we're physically restructuring our brain and creating new ways of responding to challenges.

STRATEGIC PATTERN INTERRUPTION

A "pattern interrupt," simply put, is anything that breaks an automatic thought or behavior pattern and creates a moment of pause where we can choose a different response. I first encountered this concept in my business journey a decade ago when I was learning about marketing and found out that certain colors, images, or words could interrupt consumer behavior patterns. I wondered if the same principles could be applied to my self-love journey.

I discovered that, just as an unexpected image might make someone pause while scrolling through social media, simple interruptions could create powerful pauses in my habitual thought patterns. It's like pressing a reset button on our automated programs, creating a moment of possibility where new choices can be made.

As I delved deeper into Dispenza's work, I realized that he teaches this exact technique, using the word "change" as a pattern interrupt. This validation of the approach inspired me to incorporate it into my own practice. When I noticed myself sliding into old patterns of self-doubt or criticism, mentally saying, *Change* would create an immediate pause—a moment of awareness where I could choose a different thought or response. What began as a conscious practice evolved into an automatic response, as natural as pressing pause on a video when the phone rings.

Other small but powerful pattern interrupts became woven into my environment:

- a meaningful crystal on my desk that caught my eye during stressful moments
- a specific essential oil I'd reach for when needing to center myself
- a photo on my phone that reminded me of my worth
- a gentle physical gesture that helped me return to presence

These weren't practices I had to remember to do—they had become natural responses to life's moments, as automatic as reaching for water when thirsty. Each interrupt created a moment of choice, a pause between old conditioning and new possibilities.

To bring this concept to life, I created a simple framework for pattern interruption and rewiring, which you'll see in Figure 3.2. This framework demonstrates how a trigger event—whether external or internal—activates an automatic response, which is often rooted in old, habitual programs. By introducing awareness and consciously interrupting the pattern, we create a powerful pause. This pause is where transformation happens: It provides the opportunity to reframe the old program into a new, empowering one.

Figure 3.2: Thought pattern interrupt: a neural rewiring framework

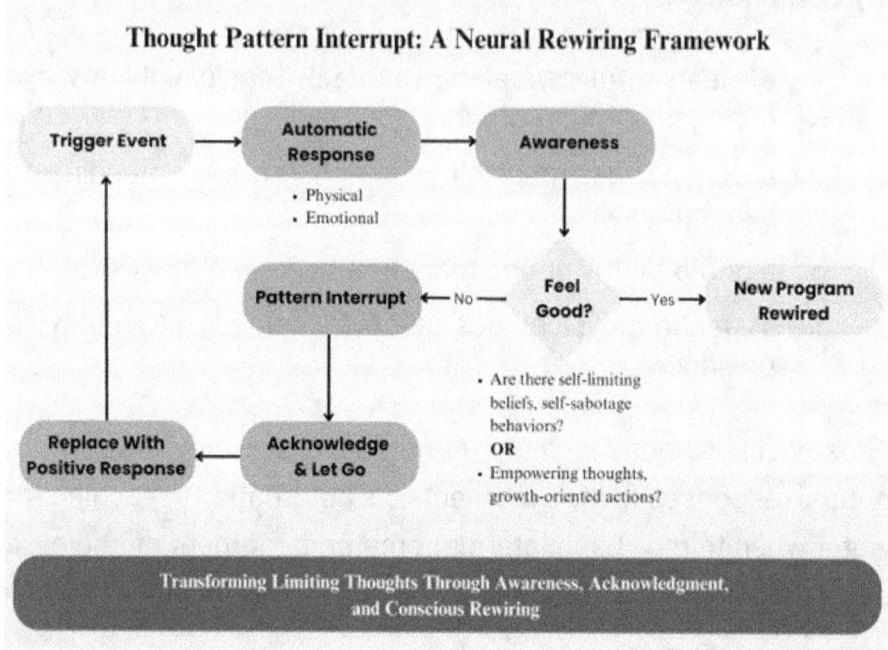

The key to this method is consistency—using it as soon as your awareness clocks that a trigger has activated an old program. This might feel like anxiety tightening in your belly, looping negative self-talk, or even noticing that you're avoiding something you know is in your best interest. These moments are invitations to interrupt the pattern, pause, and consciously choose a new response. Over time, with regular practice, this process will help you replace self-limiting beliefs with empowering thoughts and actions, ultimately rewiring your brain for lasting self-love and growth.

THE ROLE OF EMOTIONS IN REWIRING

While pattern interruption gives us the cognitive tools to break old habits, its effectiveness is amplified when we understand how our emotions participate in this process. Each thought pattern we interrupt isn't just a mental event—it carries an emotional charge that affects our entire body. Understanding this mind–body connection brings us to an even deeper level of transformation.

Thoughts and emotions are deeply interconnected, and nowhere did I experience this more powerfully than in the high-pressure moments of running my business. In his video teachings, Dispenza emphasizes that our thoughts don't just exist in our heads—they carry an electromagnetic charge, while our emotions create a magnetic pull. Together, they form what he calls our "energetic signature."

Let me share a specific example. During the early days of scaling my business, I faced a critical decision about taking on a substantial inventory investment. The opportunity was solid, but the financial commitment was significant. As soon as I began to consider the decision, my body went into its familiar stress response—tight chest, racing thoughts, shallow breathing. This physical reaction wasn't just uncomfortable; it impaired my ability to make clear-headed business decisions.

Through my study of Dispenza's work on neuroplasticity and the HeartMath Institute's groundbreaking research, I learned about the crucial relationship between heart coherence and brain function. According to the HeartMath Institute (2016), the heart generates an electrical field about 60 times greater in amplitude than that of the brain and an electromagnetic field roughly 5,000 times stronger than the brain's. This field extends approximately three feet outside our physical body and can be detected by other people's nervous systems. When we align our heart and brain through elevated emotional states, we create what HeartMath's Research Director, Dr. Rollin McCraty, defines as coherence: "the state when the heart, mind, and emotions are in energetic alignment and cooperation."

This understanding revolutionized my approach to transformation. Research shows that positive emotions like compassion and love generate harmonious patterns in the heart's rhythm, leading to greater emotional regulation and enhanced cognitive function. When we achieve this state of coherence, we can access higher cognitive functions, improve our learning capacity, and make better choices. It wasn't just about *thinking* differently; I needed to *feel* differently and allow that feeling to create physiological change throughout my body.

Dispenza teaches the concept of creating heart-brain coherence, a crucial element of this process. Rather than focusing just on visualizing success, he emphasizes the importance of feeling the emotions of your future self in your heart and connecting this with your conscious awareness. Bringing the emotions of who we want to become into our present-moment experience creates a more powerful foundation for change.

THE POWER OF FEELING YOUR FUTURE NOW

One of the most transformative insights I gained from Dispenza's work was the understanding that simply imagining or visualizing your future self isn't enough. You must feel it in your body as if it's happening now. This isn't just positive thinking; it's a biological process that creates real changes in your brain and body.

Think about this: Your brain and body don't distinguish between a vividly imagined experience and a real one. When you genuinely feel the emotions of your future self—the confidence, self-acceptance, and inner peace—your body begins producing the same chemical responses as if those circumstances were already real. You're not just dreaming about change; you're programming your neural pathways and cellular memory with the emotional signature of your future self.

As Dispenza explains in his teachings, everything you want in life, you want for the emotional payoff. Whether it's success, love, or inner peace, what you're really seeking is how you believe you'll feel when you achieve it. The revolutionary approach is to generate and embody those elevated emotions now rather than waiting for your external circumstances to change.

I discovered this firsthand during my morning practices. Instead of just visualizing myself as a confident business owner, I began to feel what that confidence felt like in my body. I would sit quietly and allow my heart to experience the deep sense of capability, wisdom, and self-trust I associated with my future self. At first, it felt artificial, almost like I was pretending. But as I continued the practice, something remarkable happened: My body began recognizing and remembering these elevated emotional states.

This process works because you're not just thinking about your future—you're creating new neural pathways that actively support your desired reality. Your brain and body begin operating as if your desired future is already present. Each time you practice feeling these elevated emotions, you're strengthening these pathways, which supports your transformation. What's even more remarkable, which we'll explore in Chapter 8, is that this process can also influence genetic expression in your body—but for now, let's focus on how it reshapes your brain's neural networks.

Here's how to deepen this practice:

1. Begin by identifying the specific emotional states you associate with your future self. Is it deep inner peace? Unshakeable confidence? Pure self-acceptance? Get clear on these feelings.

2. Instead of just thinking about these emotions, begin to generate them in your body. Ask yourself: *How would my breath feel if I were already living in complete self-love? How would my posture change? What would the rhythm of my heartbeat be?*

3. Allow yourself to be moved by these emotions. Your body might naturally want to smile, sit taller, or breathe more deeply. This physical engagement helps cement the emotional state.

4. Stay with these feelings, letting them expand through your body. Remember, you're not pretending—you're programming your nervous system with new emotional patterns.

5. Practice this, especially when you're facing challenges. If your old self-doubt arises, pause and consciously bring in the emotional state of your future self, who has already moved beyond these limitations.

Dispenza emphasizes that this isn't about forcing or faking emotions but rather about teaching your body a new way of being. When you consistently practice feeling what your future would feel like, you begin to pull that future into your present-moment experience. Your brain and body literally begin to change in response, creating new neural pathways that support your transformation.

Remember, your body is remarkably adaptable. Just as it learned to hold patterns of stress or self-doubt, it can learn to embody self-love and confidence. The key is consistency in generating and sustaining these elevated emotional states. Each time you practice, you're not just imagining a better future—you're literally rewiring your brain and reprogramming your body to manifest that future.

This isn't about pretending or forcing positive feelings. It's about actually experiencing the emotional state of your future self now, allowing your brain and body to begin recognizing this as your new normal. The more you practice this, the more naturally these states will become available to you in your daily life.

☆ PRACTICAL EXERCISE: FUTURE SELF EMBODIMENT

This transformative practice helps you bring the emotions of your future self into your present moment. Remember, this isn't just visualization—it's about creating real, physiological changes in your brain and body through elevated emotional states.

STEP 1

Find a quiet space where you won't be interrupted. Take a few deep breaths to center yourself. As you breathe, feel yourself settling deeper into your body. Let each exhale release any tension or resistance to change.

STEP 2

Begin by focusing on your heart space. Slow and steady your breath. With each breath, imagine your heart's electromagnetic field expanding, creating a sphere of coherent energy around you. Feel this field becoming stronger and more stable with each breath.

STEP 3

Create a clear vision of your future self—the version of you who has already developed deep self-love and inner confidence. This could be

- the you who maintains healthy boundaries effortlessly, feeling peaceful and empowered in every interaction.
- the you who feels deeply worthy and loved in your relationships, radiating authentic confidence.
- the you who moves through life with unshakeable self-trust, making decisions with clarity and ease.
- the you who radiates self-acceptance and inner peace, inspiring others through your presence.

STEP 4

Now, deeply engage with this future self, using all your senses. Notice

- how they carry themselves—feel the strength and grace in their posture.
- the look of peace in their eyes—sense the deep well of wisdom and self-acceptance behind their gaze.

- the warmth they radiate—feel how their presence naturally draws others toward them.
- the confident smile that comes effortlessly to them—experience the genuine joy and ease in their expression.
- the way they move—feel the groundedness and fluidity in each of their steps.

STEP 5

This is the crucial step, where you fully embody these feelings in your present moment. Feel the emotions of this future self as if they're happening right now. Let them permeate every cell of your body:

- feel the deep sense of self-worth settling into your bones
- experience the unwavering self-trust anchoring in your solar plexus
- let profound inner peace flow through your nervous system
- allow natural self-acceptance to soften every muscle

Notice how these elevated emotions create physical sensations in your body, perhaps in the following ways:

- a warmth spreading through your chest
- a relaxing of your shoulders
- a natural straightening of your spine
- a deepening of your breath
- a slight upturning of your lips into a gentle smile

STEP 6

Let these feelings expand in your heart. Feel them becoming stronger and more real. Remember, your brain can't distinguish between vividly imagined emotions and real ones. As you maintain these elevated states

- notice how your heart rate might shift.
- feel the coherence building between your heart and brain.
- experience how your body relaxes into these new emotional patterns.
- allow waves of gratitude to amplify these positive states.

STEP 7

As you prepare to end the practice, consciously integrate these elevated emotions into your being. Feel them becoming part of your cellular memory, creating coherence between your heart and brain. Take a moment to appreciate how different your body feels now compared to when you started.

Before opening your eyes, commit to carrying these elevated emotions with you. Remember that each time you practice this embodiment

- you're strengthening new neural pathways.
- you're creating new emotional patterns in your body.
- you're broadcasting a new electromagnetic signature.
- you're literally becoming your future self in this present moment.

This isn't about pretending or forcing positive feelings. It's about actually experiencing the emotional state of your future self now, allowing your brain and body to begin recognizing this as your new normal. The more you practice this exercise, the more naturally these elevated states become available to you in your daily life.

THE POWER OF REPETITION

While creating a vivid vision of your future self is powerful, the real transformation happens through consistent practice. This is where many people get discouraged—they try something once or twice and give up when they don't see immediate results. As Dispenza teaches, rewiring your brain takes time and consistency. The science shows us that repeatedly activating these new neural pathways is what makes them stronger.

> *"Neurons that fire together, wire together."*
> **—Joe Dispenza**

The more you practice self-love, the more natural it becomes. Think about it: Your current thought patterns didn't develop overnight. They're the result of years of repetition. Similarly, building new self-love patterns requires regular practice. It's like learning any new skill—at first, it feels awkward and requires conscious effort, but with repetition, it becomes your natural way of being.

MAKING IT PART OF YOUR DAILY LIFE

The key is to integrate these practices into your daily routine in a way that feels manageable and sustainable. Consistency is crucial for rewiring neural pathways, so I've created a simple tracking system to help you maintain your practice throughout the week.

Here's how to structure your daily practice:

- **Morning visualization:** Begin with the "future self" visualization we just learned. Even 10 minutes of consciously embodying your future self sets a powerful tone for your day.
- **Midday check-in:** Take a brief moment to reconnect with the elevated emotions you cultivated in the morning. This helps maintain the coherent state and strengthens the new neural pathways you're creating.
- **Evening reflection:** Before bed, acknowledge your growth and the moments when you successfully accessed these new patterns. This reinforcement helps consolidate the day's progress.

Use a tracker to mark each practice you complete. Don't aim for perfection—aim for progress. Even if you miss a session, simply begin again at the next opportunity. Remember, every time you engage in these practices, you're physically restructuring your neural pathways and strengthening your capacity for self-love.

Here's a simple tracker you can choose to use to help you maintain consistency:

	Morning visualization	Midday check-in	Evening reflection
Monday	✓	✓	
Tuesday	✓		✓
Wednesday	✓	✓	✓
Thursday	✓		✓

📌 KEY TAKEAWAYS: REWIRING YOUR BRAIN FOR SELF-LOVE

- ➤ Your brain has the ability to create new neural pathways through neuroplasticity. Old patterns aren't permanent; they can be rewired.
- ➤ Combining elevated emotions with new thought patterns creates stronger, more lasting changes in your brain.
- ➤ Bringing the emotions of your future self into the present moment is more powerful than positive thinking alone.
- ➤ Creating heart–brain coherence amplifies your ability to establish new self-love patterns.
- ➤ Consistency and repetition are essential—you strengthen these new neural pathways each time you practice them.
- ➤ The goal isn't perfection in practice but persistence in creating new patterns.
- ➤ Your physical health and emotional well-being are directly connected to these thought patterns and neural pathways.

Remember: Change happens through conscious, consistent practice. Whenever you choose a new thought or emotion, you're physically reshaping your brain's architecture.

The mind, with its thoughts,
IS DRIVEN BY FEELINGS
Each feeling is the
cumulative derivative
of many thousands of thoughts.
Because most people
throughout their lives
*repress, suppress, and try to
escape from their feelings,*
the suppressed energy accumulates
and seeks expression
through psychosomatic distress,
bodily disorders, emotional illnesses,
and disordered behavior
in interpersonal relationships.

The accumulated feelings
block spiritual growth and awareness,
as well as success in many areas of life.

DAVID R. HAWKINS

CHAPTER 4:

Releasing Emotional Blocks to Self-Love

For most of my life, I was carrying emotional blocks from childhood trauma that were so deeply suppressed and repressed, I didn't even know they existed. These emotions—guilt, shame, anger, and fear, which I later learned through Hawkins' work were the lowest vibrational frequencies—weren't obvious or visible to me. They were buried so deep beneath the surface that I had no conscious awareness of them, yet their presence required enormous amounts of energy for me to function in daily life. Simple acts of courage or showing up in the world demanded so much more effort than they should have done, all because I was unconsciously carrying the weight of these emotional blocks. It was like trying to swim while wearing heavy clothes: I was expending so much energy to stay afloat, but I couldn't identify what was weighing me down.

Through my work in this area, I came to understand that many of these stored emotions weren't just memories; they were carrying the energy of my younger self, my inner child, who didn't have the resources or ability to process these experiences at the time. This part of me had learned to suppress these emotions as a survival mechanism because, as a child, I didn't have the tools, support, or safety to express and release them.

WHAT ARE EMOTIONAL BLOCKS?

Through my study of Singer's work, I discovered that emotional blocks are unresolved energy patterns stored within us. In his book *The Untethered Soul*, Singer explains that these blocks—known as *samskaras* in Indian philosophy—are like emotional "scar tissues." What fascinated me was learning that these impressions aren't just emotional memories; they're stored energy in our system that continues to affect us until we release it.

The process of developing these blocks often starts in childhood. In my case, early trauma led to the suppression of emotions that felt too overwhelming to process at the time. Instead of naturally experiencing and releasing these feelings, I unconsciously stored them. Singer explains that when you resist an emotion, you don't let it pass through you. Instead, it becomes locked in your system and can be triggered years later, forcing you to relive it as if it's happening all over again.

I then learned that from a neuroscience perspective, these unresolved emotions are like closed loops in the brain. They keep firing the same neural circuits, reinforcing negative thought patterns. Until we release these blocks, our brain continues to run the old programs, making it challenging to create new, healthier patterns of thinking and feeling.

What made this understanding so impactful for me was realizing that these stored emotions weren't just affecting my mood or relationships—they were consuming vast amounts of my life-force energy. Every day, I was unconsciously using enormous amounts of energy to suppress these emotions, leaving me with less energy for growth, creativity, and joy.

It's like running a computer with dozens of hidden programs constantly running in the background. They don't just drain the battery; they eventually affect the performance of the entire system. Just as an overloaded computer shows signs of strain—freezing, overheating, or crashing—our bodies manifest physical symptoms when we suppress too many emotions. This takes a toll not just on our energy levels but also on our physical health and well-being.

Hawkins articulates this connection between suppressed emotions and physical health powerfully in his book *Letting Go: The Pathway of Surrender*.

> *"Chronic, unrecognized anger and resentment re-emerge in our life as depression, which is anger directed against oneself. If pushed further into the unconscious, it can re-emerge as a psychosomatic illness. Migraine headaches, arthritis, and hypertension are frequently cited examples of chronic suppressed anger."*
>
> **—David R. Hawkins, *Letting Go: The Pathway of Surrender***

This hit home for me personally, as I had been suffering from chronic migraines since my early teens. I hadn't made the connection initially, but as I began this work of emotional release, I started to understand that my migraines were a physical manifestation of my suppressed emotions. My body was speaking to me through pain, trying to get my attention about what needed to be released.

This revelation helped explain why certain situations that seemed simple for others felt so exhausting for me. What I thought was just "who I am" was my system working overtime to maintain these suppressions. Singer's insight aligned perfectly with Hawkins' teaching about emotional frequencies. These suppressed emotions of guilt, shame, anger, and fear vibrated at the lowest frequencies, requiring constant energy to maintain their burial. The path to releasing these blocks began with a crucial first step: acceptance.

One of the most powerful examples of this in my life came through a relationship experience. When I developed strong feelings for someone who didn't reciprocate them, my initial response was resistance—fighting against the reality of the situation, trying to force a different outcome. The pain of this unrequited love triggered all my stored emotions of unworthiness and fear. However, through applying Singer's teachings about acceptance, something remarkable happened. As I learned to accept the situation exactly as it was, without trying to change it, I discovered I could love unconditionally without expecting anything in return. This was a powerful lesson in emotional freedom.

THE POWER OF ACCEPTANCE

That first experience of acceptance in the face of unrequited love became a gateway to a much deeper understanding. What began as accepting one particular situation evolved into a profound life practice. I started to recognize how this same principle of acceptance could be applied to other areas of my life where I had been carrying resistance, resentment, guilt, and shame.

And here's something else that's vital to understand: Acceptance doesn't mean tolerating toxic situations or abandoning healthy boundaries. In fact, true acceptance often gives you the clarity and strength to establish better boundaries. You can accept a situation for what it is while still choosing to protect your energy and well-being. For instance, you might accept that someone has toxic behavioral patterns while simultaneously choosing to limit your interaction with them. This is acceptance paired with self-respect and healthy boundaries.

Once I'd experienced acceptance in my body, heart, and energy field, something remarkable happened. I discovered I could feel the difference in my vibrational frequency when I was in acceptance versus when I was in resistance. This bodily awareness became an internal compass. Whenever I noticed myself tensing against life—whether it was a business challenge, a relationship issue, or an old memory surfacing—I could consciously choose to return to the state of acceptance I had come to know.

This practice led me to another significant understanding: Acceptance and trust are intimately connected. When you truly accept what is, you open yourself to trusting that whatever is presenting itself in your life has a purpose. This was revolutionary for me. Instead of seeing challenges as something to fight against, I began to view them as opportunities for growth and release.

For example, when old emotional blocks would surface—perhaps triggered by a current situation—instead of trying to suppress them again or fight against feeling them, I learned to say, *Okay, this is here right now. I accept its presence.* This acceptance allowed these emotions to finally move through me instead of staying stuck. It was like finally opening a door that had been locked for years, allowing stagnant energy to flow out.

The practice followed this path: noticing resistance, choosing acceptance, and trusting the process. With each application of this approach, I could release layers of old emotional blocks that had been weighing me down for decades. The energy involved in maintaining these blocks gradually became available for living fully in the present moment.

HOW EMOTIONAL BLOCKS SHOW UP IN DAILY LIFE

While my journey with emotional blocks manifested primarily through physical symptoms like migraines and energy depletion, these blocks can show up differently for different people. Recognizing how these blocks manifest in your life is the first step toward releasing them.

Here are some common ways emotional blocks might be showing up in your life:

- **Self-criticism:** You might feel you are "not enough"—not smart enough, not successful enough, not attractive enough. These persistent feelings of inadequacy often stem from unprocessed experiences of failure or rejection. For instance, you might nail a presentation at work but still focus on the one small mistake you made, or you could receive mostly positive feedback but obsess over the one critical comment.
- **Sabotaging relationships:** Emotional blocks often surface most clearly in relationships. For example, fear of abandonment might cause you to push people away before they can leave you, or you might find yourself

staying in unfulfilling relationships because they feel "safe." While understandable, these self-protective behaviors ultimately damage your ability to form genuine connections with others.

- **Difficulty accepting compliments:** Pay attention to how you respond when someone offers you praise or appreciation. Suppose you immediately dismiss compliments, make self-deprecating jokes, or feel uncomfortable with positive attention; this might indicate blocks around worthiness and self-value. This pattern reinforces a cycle of low self-esteem by literally pushing away evidence of your worth.

- **Avoiding self-care:** One of the most telling signs of emotional blocks is discomfort with self-care. You might find yourself

    » feeling guilty about taking time for yourself.
    » struggling to set boundaries, even when you're overwhelmed.
    » putting everyone else's needs before your own.
    » viewing self-care as selfish or indulgent.

These patterns often stem from deep-seated beliefs about your right to prioritize your well-being.

RELEASING THROUGH CONSCIOUS AWARENESS

Whether your emotional blocks manifest as self-criticism, relationship patterns, health issues, or difficulty with self-care, the path to freedom begins with acceptance. Just as acceptance opened the door to transformation in my journey, it's the key that will unlock your ability to work with these blocks, whatever their form.

It's important to note that acceptance doesn't mean resigning yourself to these patterns or letting go of boundaries. Instead, it creates the space needed for genuine change. When we accept how these blocks show up in our lives, without judgment or resistance, we can begin releasing them.

Studying Hawkins' and Singer's work, I discovered that emotional release happens through awareness, acceptance, and consciously letting go. Their teachings and my experience revealed this powerful truth: No emotional block is permanent when we approach it with acceptance and understanding.

The process begins with being willing to fully experience our suppressed emotions—the guilt, shame, anger, and fear that we've worked so hard to avoid. This might sound counterintuitive or even frightening at first. After all, many of us have spent years trying to keep these uncomfortable emotions at bay. But here's what I learned: Suppressing emotions requires enormous energy. When we finally allow ourselves to feel them in a safe, conscious way, we begin to free up that energy. It's like finally unplugging all those background programs draining our battery.

☆ THE PRACTICE OF EMOTIONAL RELEASE

While you can set aside dedicated time for emotional release work, I discovered that one of the most powerful opportunities for this practice occurs when emotions naturally arise. When you receive that triggering email, when someone says something that sparks an emotional response, or when you're in a challenging situation—these are optimal moments to practice this work.

Here's the process that worked for me.

CATCH IT EARLY

When you notice an emotional response beginning—perhaps your stomach tightens after reading an email or your chest constricts during a difficult conversation—this is your cue to begin the practice.

STEP 1: CREATE A MINI CONTAINER

While you might not be able to find a tranquil space at the moment, you can create a brief pause. This might mean

- stepping away from your desk for a moment.
- taking a brief bathroom break.
- sitting in your car for an extra minute.
- simply closing your eyes at your desk.

STEP 2: LOCATE AND ACKNOWLEDGE

Different emotions tend to show up in specific areas of the body. For me

- anger often manifested in my lower body and pelvic area.
- shame and guilt frequently showed up in my upper abdomen.
- fear typically appeared as a tightness in my chest.
- anxiety usually created tension in my throat.

STEP 3: CONNECT WITH YOUR INNER CHILD

If you recognize that your emotional response is related to your past experiences, especially childhood, take a moment to acknowledge your inner child. I often found myself saying something like, *I understand why you felt this way then. Those feelings were completely valid. You did what you needed to do to survive. But adult me is here now, and we're safe.*

STEP 4: ALLOW WITH COMPASSION

Rather than trying to figure out why you're triggered or whether your reaction is justified, allow the emotion to be there with gentleness. Remember, you're not just accepting the emotion—you're creating a safe space for that younger part of you to finally release what it's been holding.

1. **Acknowledge your adult resources:** Remind yourself that you have resources now that you didn't have as a child. For example, you can say to yourself, *I know how to handle this situation now. I can communicate my needs, set boundaries, and keep us safe.*
2. **Allow it to flow through:** This is where Singer's teaching becomes particularly powerful. Once you've acknowledged the emotion and connected with your inner child, consciously allow the energy to pass through you rather than storing it in your body. Imagine yourself becoming completely permeable, letting the emotion flow like water through a sieve. Don't try to hold onto it or push it away; allow it to move.

This process of letting emotions pass through might

- take just a few moments during a brief pause in your day.
- require several minutes of conscious breathing.
- call for a more extended period of meditation or quiet reflection.
- unfold gradually over multiple sessions.

The key is to take your time with the process. Some emotions, especially deeply stored ones, might need multiple opportunities to be fully released. You might find that

- the intensity of the emotion moves up and down.
- physical sensations shift or move through different parts of your body.
- spontaneous sighs or yawns occur as the energy releases.
- a natural sense of relief or lightness emerges.

STEP 5: SUPPORT THE RELEASE

You can support this flow-through process with

- deep, conscious breathing.
- gentle movement or stretching.
- quiet meditation.
- focused breathwork.

Hawkins and Singer emphasize that our natural state is one of flow and release. We're not trying to create something new—we're simply returning to our natural ability to let emotions move through us rather than get stored in our bodies.

The more you practice this, the more you'll trust your body's innate wisdom to process and release emotions. What once might have taken days of emotional turmoil can often be released in moments when you fully allow it to pass through.

WHAT HAPPENS AFTER RELEASE

When you begin practicing emotional release regularly, you might notice immediate and gradual changes in how you experience life. For me, the shifts happened on multiple levels.

The immediate effects may include

- a physical sensation of lightness, as if a weight has been lifted.
- deeper, more natural breathing.
- a sense of spaciousness where tension used to be.
- mental clarity that wasn't there before.
- a natural feeling of calm or peace.

Sometimes, you might also experience

- temporary fatigue as your system integrates the release.
- spontaneous yawning or sighing.
- a need for quiet or rest.
- emotional sensitivity.

These are all typical signs that energy is moving and shifting in your system. Think of it as an emotional detoxification process.

Over time, with consistent practice, you might notice

- greater emotional resilience.
- more energy available for daily life.
- a natural ability to let emotions flow rather than get stuck.
- increased trust in your ability to handle emotional challenges.
- a deeper connection with yourself.
- more authentic relationships with others.
- a growing sense of inner peace and self-love.

The beauty of this practice is that each release creates more space for your natural state of well-being to emerge. As Singer explains, you're not adding anything new—you're simply removing the blocks that have been preventing your natural state of peace and self-love from expressing itself.

MOVING FORWARD WITH FREEDOM

As you begin to work with these awareness, acceptance, and release practices, remember that this is a journey of returning to your natural state. Every emotion you allow to flow through you, every block you release, creates more space for your authentic self to emerge. The energy once consumed by suppressing these emotions will become available for living fully and loving yourself deeply.

This work can be challenging. There will be days when the emotions feel intense or old patterns try to reassert themselves. But with each practice session, whether it's a momentary pause to process a trigger or a more extended period of conscious release, you're rebuilding your natural capacity to let emotions flow through you rather than them getting stored in your system.

Remember, your body knows how to do this. Just as it knows how to heal a cut or fight an infection, it knows how to process and release emotions. Your job is to create the conditions—through awareness, acceptance, and conscious release—that allow this natural process to occur.

📌 KEY TAKEAWAYS: RELEASING EMOTIONAL BLOCKS

- Emotional blocks are stored energy patterns that consume your life force and can manifest as physical symptoms when suppressed.

- Acceptance is an active, powerful choice that creates the space for emotional release while maintaining healthy boundaries.

- Your inner child often holds these emotional blocks, and healing involves creating safety for these younger parts of yourself.

- Physical symptoms like chronic pain or illness may be manifestations of suppressed emotions seeking release.

- Emotional release is a natural process that happens when you allow yourself to feel and let emotions pass through you rather than storing them.

- Real-time processing of emotions as they arise can prevent new blocks from forming.

- Consistent practice of emotional release creates more energy, better health, and a greater capacity for self-love.

Remember: Each emotion you release creates more space for love to flow naturally in your life.

The wound of the past
finds healing in the
TENDER EMBRACE

of present awareness.

FIONA SOUTTER

CHAPTER 5:

Embracing Your Inner Child

The moment I truly understood the power of connecting with my inner child remains vivid in my memory. I was walking from my office toward the kitchen when a wave of fear and unease washed over me regarding a business situation. As I reached the doorway, I paused, turning my awareness inward. In that moment of stillness, I recognized that this fear wasn't coming from my adult self—it was "Little Fe," the younger part of me that still carried old patterns of uncertainty and self-doubt.

Instead of pushing the feeling aside or trying to power through it, I chose to have a gentle internal dialogue. *It's okay*, I said silently to my younger self. *I understand why you feel this way, but everything's okay now. I know what to do.* I recognized that this trigger, which had once served as a vital protection mechanism in my youth, had outlived its purpose. What had been a shield for my younger self had become an unnecessary barrier in my adult life.

After this acknowledgment, I consciously chose to release the trigger from my body, understanding that I no longer needed this old survival response. I felt the fear and tension flow out as I walked to the kitchen, like letting go of a heavy coat I'd been wearing in warm weather. This wasn't just positive self-talk—it was a heartfelt moment of acknowledging and nurturing my inner child while maintaining my adult presence and wisdom, followed by intentional energetic release and the recognition that I had outgrown this particular defense mechanism.

After getting a drink and returning to my office, I noticed a complete shift in my energy and confidence regarding the situation that had triggered fear and unworthiness moments before. This experience showed me the power of recognizing and separating my younger self's emotions from my adult capabilities while also honoring the importance of physical and energetic release. It was a transformative moment of deep self-love—one that demonstrated how nurturing my inner child and consciously releasing old patterns could lead to radical shifts in my present-day experience.

THE NEUROSCIENCE BEHIND INNER CHILD PATTERNS

My research revealed that these experiences weren't just psychological insights—they had a clear neurobiological basis. A groundbreaking study published in *Nature Communications* (Sorrells et al., 2019) discovered something remarkable: A group of cells in the amygdala—our brain's emotional processing center—sometimes remains unchanged from childhood into adulthood. Unlike other brain cells that mature and adapt as we age, these particular neurons in the amygdala's paralaminar nuclei can remain immature, and they have even been found in this unchanged state in a 77-year-old brain. This finding provides intriguing neurobiological support for what many of us experience—the presence of childhood emotional patterns persisting into adulthood.

Through my work with Dr. Espen, I learned that childhood experiences create particularly strong neural imprints because our brains are most plastic during these early years. The emotions we couldn't fully process as children—whether they stemmed from significant trauma or seemingly small moments of feeling unsafe, unloved, or unseen—become stored in our neural pathways and cellular memory. Our inner child holds these emotional patterns in trust, carrying them forward until we're ready and able to address them as adults.

WHEN YOUR INNER CHILD EMERGES

The neural patterns we develop in childhood surface in specific situations that echo past experiences or trigger old survival patterns. While that moment in my office doorway was a clear recognition of "Little Fe," I began noticing her presence in various aspects of my life:

- during challenging business negotiations, where speaking up for my worth would trigger old fears of rejection
- in relationships, where setting boundaries felt frightening because doing so hadn't felt safe in childhood
- when facing uncertainty, particularly around financial decisions, which brought up childhood anxieties about security
- in moments of conflict, when the urge to people-please and make myself smaller would emerge.

What made this understanding so powerful was recognizing that these weren't character flaws or signs of weakness—they were moments when my inner child's protective patterns were being activated. This understanding transformed how I approached these moments of activation. Instead of getting frustrated with myself for feeling anxious or uncertain in situations where my adult self knew I was capable, I could recognize these as opportunities to

- acknowledge my inner child's presence and legitimate fears.
- provide the safety and reassurance that the younger part of me needed.
- access my adult resources and wisdom.
- consciously release the triggered energy from my body.
- move forward with integrated strength.

I experienced this personally through an automatic response that frustrated me for years—tears that would spring up during constructive criticism or when someone expressed disappointment in me. As an adult, these tears felt inappropriate and overwhelming. I would often get annoyed with myself, trying to suppress them while thinking, *Why am I crying? There's no need for this!*

What I didn't realize then was that by being frustrated with these tears and trying to suppress them, I was being unkind to my inner child, who'd had very real reasons to cry in similar situations during childhood. There could be serious consequences for that little girl when she didn't meet expectations or do the right thing. Her tears weren't an overreaction but a natural response to an environment where being "good" was tied to her safety and survival.

Understanding this transformed how I responded to these moments when they arose. Instead of fighting against the tears or feeling frustrated with myself, I learned to pause and acknowledge my inner child's response with compassion. This didn't mean I needed to stay in that emotional state—quite the opposite. By recognizing these tears as a signal from my younger self, I could consciously choose to comfort that part of me while also accessing my adult capacity to handle the present situation appropriately.

I understand why you feel this way, I would tell my inner child, *but we're safe now, and adult Fiona knows how to handle this*.

Each time I acknowledged and released these patterns with love, I strengthened my adult presence while honoring my inner child's experience. This wasn't about suppressing or "fixing" my inner child's responses—it was about creating a new, more empowering relationship between my adult self and my younger self. Each time I noticed "Little Fe" emerging, it became an opportunity to practice this deeper form of self-love and integration.

☆ TOOLS FOR INNER CHILD CONNECTION AND HEALING

After understanding the deep connection between my adult triggers and my stored childhood experiences, I needed practical ways to work with these moments as they arose. The tools I'm about to share emerged from years of personal practice and professional study. They're designed not just to help you cope with emotional triggers but also to create lasting transformation through conscious, compassionate engagement with your inner child.

One of the most powerful tools I developed was what I call the "conscious dialogue practice." This isn't just about talking to yourself—it's about creating a safe space for your inner child while maintaining your adult presence. Here's how it works.

STEP 1: RECOGNITION AND PAUSE

When you notice emotional patterns or triggers arising, take the following steps:

1. Stop what you're doing (like my moment in the doorway).
2. Take a conscious breath.
3. Turn your awareness inward.
4. Notice where in your body you feel the emotion or sensation.

STEP 2: ACKNOWLEDGMENT

1. Recognize this as your inner child's response.
2. Validate whatever feelings are present without judgment.
3. Understand that these feelings arose from how you interpreted situations through a child's perspective. For example, not being chosen for a team might have been interpreted as "I'm not good enough," or a busy parent saying "not now" might have been felt as "I'm not important."

STEP 3: ADULT SELF CONNECTION

- Consciously shift into your adult perspective.
- Access your current resources, wisdom, and understanding.

- Remember you have the capacity now to hold space for these feelings while knowing your true worth.

STEP 4: GENTLE DIALOGUE

Use phrases that combine acknowledgment with reassurance:

- I understand why you feel [scared/unworthy/unloved/unsafe/hurt/not enough]. Those feelings made sense from your perspective then.
- I'm here now, and I know how to navigate this.
- You're safe and loved. Adult [your name] has this covered.
- I see you and understand what you're feeling. We can handle this together now.
- It's okay to have these feelings, and it's also okay to let them go.

STEP 5: CONSCIOUS RELEASE

- Intentionally release the triggered energy from your body.
- Allow any emotions present to flow through you.
- Let your body naturally reset to a state of calm.
- Trust in your adult capability to move forward with clarity.

This practice becomes particularly powerful when combined with the understanding of neuroplasticity we discussed earlier. Each time you engage in this dialogue while maintaining your adult presence, you create new neural pathways that integrate your inner child's experiences with your adult wisdom and capabilities.

LIVING WITH YOUR INNER CHILD: AN ONGOING JOURNEY

Understanding and working with your inner child isn't a destination; it's an evolving relationship that continues to deepen over time. Through consistent practice, I discovered that what began as conscious effort gradually transformed into natural wisdom. I became more attuned to the moments when my inner child needed attention and could respond with grace and understanding.

This evolution meant that my inner child work became less about managing "triggers" or "fixing" responses and more about maintaining an ongoing dialogue of love and understanding between my adult self and younger aspects. Sometimes, this meant celebrating moments of joy and playfulness. Other times, it meant providing comfort and reassurance during challenges. It always meant maintaining that crucial balance between acknowledging my past experiences while staying grounded in my adult capacity to handle present situations.

The most beautiful part of this journey has been witnessing how this relationship with my inner child has enhanced rather than hindered my adult capabilities. Each time I acknowledged and released old patterns with love, I strengthened my adult presence. Understanding and caring for the part of me that feared speaking up made me more confident in expressing myself authentically. My inner child became not just a part to be healed but a source of creativity, wonder, and emotional depth that enriched my adult life.

EMBRACING THE JOURNEY WITH YOUR INNER CHILD

Your inner child isn't just a part of your past—it's an active participant in your present journey of growth and self-discovery. When you learn to dance between your adult wisdom and your inner child's truth, you create a fuller, richer experience of self-love. On some days, this might mean taking a moment to comfort your inner child during a challenging meeting. On other days, it might mean allowing your inner child to express joy freely when something delights you. Each interaction becomes an opportunity to strengthen both your adult presence and your connection to your younger self.

As you move forward, trust that this relationship will continue to evolve naturally. Just as any meaningful relationship deepens over time, your connection with your inner child will grow richer through each moment of acknowledgment, each gesture of comfort, and each conscious choice to create safety for all parts of yourself.

📌 KEY TAKEAWAYS: EMBRACING YOUR INNER CHILD

- Your inner child's responses have a neurobiological basis, with some of your emotional processing neurons remaining unchanged from childhood.
- Creating a compassionate relationship with your inner child allows both the healing of past wounds and the integration of emotions in the present.
- Conscious dialogue between your adult self and your inner child creates new patterns that support emotional regulation and self-love.
- Regular practice isn't about perfection—it's about creating sustainable patterns of self-understanding and nurturing.
- Your inner child carries not only past experiences but also your capacity for joy, wonder, and authentic expression.
- Each interaction with your inner child is an opportunity for both healing and growth, allowing all aspects of yourself to contribute to your journey of self-love

Remember: Your inner child carries not only past hurts but also your capacity for joy, wonder, and authentic expression. As you develop this relationship with compassion and understanding, you'll create space for healing and growth, allowing all aspects of yourself to contribute to your self-love journey.

MAKE A DIFFERENCE WITH YOUR REVIEW

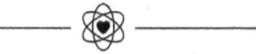

"When you light another's path, you also brighten your own."

People who help others on their journey of self-discovery create ripples of positive change.

Let's create these ripples together! Would you help someone just like you—someone who's ready to transform their relationship with themselves but doesn't know where to start?

My mission is to make the journey to self-love clear and accessible for everyone by bridging science with soul. But to reach more people who need this message, I need your help.

Most people choose books based on reviews. Your honest review could be the lighthouse that guides someone to their own transformation. Your words could help...

...one more person break free from self-doubt

...one more heart heal from past wounds

...one more soul discover their worth

...one more mind understand their power to change

...one more life transform through self-love

To make a difference, simply scan the QR code and leave a review.

Your review might be exactly what someone needs to hear to begin their own journey to self-love. Thank you for being part of this ripple of transformation!

With gratitude,

Fiona

Gratitude is the best medicine
It heals your mind,
your body, and your spirit,
and attracts more things

to be grateful for.

DAVID R. HAWKINS

CHAPTER 6:

Cultivating Gratitude as a Gateway to Self-Love

After learning to observe my thought patterns and release emotional blocks, I discovered another powerful gateway to self-love: gratitude. While gratitude had always come naturally to me, it wasn't until I began studying neuroscience that I understood its profound potential as a conscious practice.

My journey with conscious gratitude practice began in nature. I would take time to fully immerse myself in the present moment, deliberately noticing and appreciating every detail—the way sunlight filtered through leaves, the gentle touch of a breeze, the intricate patterns in tree bark. This practice of present-moment gratitude expanded to include appreciation for the people in my life, the opportunities I'd been given, and even the challenges that were helping me grow.

However, the most transformative aspect of gratitude emerged through my journaling practice. What began as simple daily reflection evolved into a powerful tool for self-discovery and healing. Through this practice, I discovered that gratitude could do more than just shift my perspective—it could fundamentally transform my relationship with myself, including those parts I'd once tried to change or hide. The simple act of finding gratitude, even in life's challenges, became a cornerstone of my self-love journey.

This wasn't just about feeling temporarily better—gratitude was creating lasting changes in my neural architecture, forging new pathways that supported a more positive, accepting relationship with myself. As I would later discover through my research, this transformation had a scientific basis, one that explained why gratitude could create such far-reaching and lasting change.

THE SCIENCE OF GRATITUDE

The power of conscious gratitude isn't just anecdotal—it's backed by neuroscience. Dispenza explains that practicing gratitude activates the brain's reward system, releasing dopamine and serotonin—our "feel-good" neurotransmitters. Dopamine, associated with pleasure and reward, creates a positive feedback loop that encourages more grateful thinking, while serotonin helps regulate mood and contributes to a sustained sense of well-being.

This neurochemical cocktail changes the structure of our neural pathways over time, making gratitude increasingly natural. I found this particularly fascinating when I discovered research showing that people who write gratitude letters, along with other personal growth practices, experience faster emotional recovery and improved well-being compared to those who journal about challenging experiences only (Wong et al., 2018). This aligned perfectly with my own experience of how combining gratitude practices with the emotional release work we discussed in Chapter 3 accelerated my transformation.

Building on this understanding, the same research has shown that gratitude practices affect multiple brain systems, from the limbic system that processes our emotions to the hypothalamus that regulates our sleep and stress responses. This explains why people who maintain regular gratitude practices often report better sleep quality and increased emotional resilience.

When we consciously practice gratitude, we rewire our brains to seek positivity and abundance. Dispenza's research and teachings show us how gratitude helps create new neural pathways that support the receipt of positive experiences and emotions. This understanding helped me grasp why my journaling practice was so effective. Whether I was feeling grateful for lessons from the past, appreciating moments in the present, or expressing gratitude for future manifestations as if they'd already happened, I was creating new neural pathways that made it easier to access positive states.

> *"Gratitude serves as a bridge, helping us move from lower emotional states into higher frequencies of acceptance and love."*
> **–David R. Hawkins**

Through his groundbreaking research, Hawkins identified gratitude as one of the highest-frequency emotions on his consciousness scale, resonating just below unconditional love. This explained why ending my journaling sessions with gratitude felt so powerful—I was elevating my emotional frequency, even when exploring emotionally challenging topics.

This practice creates what I came to think of as an upward spiral. As Dispenza explains, when we practice gratitude regularly, we strengthen the neural circuits that support positive emotional states. This makes it progressively easier to

- find the gifts in our challenges.
- appreciate aspects of ourselves we once criticized.
- access states of self-love and acceptance.
- transform our relationship with ourselves.

TRANSFORMING SELF-PERCEPTION THROUGH GRATITUDE

Understanding the science behind gratitude is powerful, but the real magic happens when we begin applying it to our relationship with ourselves. Through my journey, I discovered that gratitude could be about more than just appreciating the good things in life—it could transform how I viewed my challenges, my past, and even the parts of myself I had once tried to change.

This transformation began with my journaling practice. Unlike traditional gratitude journals, which focus solely on listing positive things, I developed an approach that helped me work with whatever was present in my life—including the difficult stuff. The key was finding something to be grateful for about that day's topic, no matter how challenging it seemed at first.

For example, when exploring a trigger that had emerged during a difficult conversation, I would write about what this trigger was teaching me about myself instead of just focusing on the discomfort. It could be showing me where I needed stronger boundaries or revealing an old pattern that was ready to be healed.

Finding gratitude for these insights transformed them from purely painful experiences into opportunities for growth.

This practice began to shift how I viewed every aspect of my journey. Remember those emotional blocks we discussed in Chapter 3? They became teachers rather than obstacles. The patterns I was working to change became witnesses to my resilience rather than evidence of failures.

Here's how this transformative gratitude practice began showing up in different areas of my life.

REFRAMING PAST EXPERIENCES

When I looked back at what I'd once viewed as failures or mistakes, gratitude helped me see the wisdom I'd gained. Those business decisions that hadn't worked out? I felt grateful for how they'd taught me to trust my intuition. Relationships that had ended painfully? I learned to appreciate how they'd shown me what I truly needed and deserved.

WORKING WITH CURRENT CHALLENGES

Rather than pushing away challenging emotions or situations, I began to ask myself: *What might I be grateful for in this moment?* When anxiety arose about a big decision, instead of fighting it, I'd find gratitude for my body's wisdom in alerting me to pay attention. When facing a difficult conversation, I'd appreciate the opportunity to practice speaking my truth.

BUILDING SELF-WORTH

The most substantial shift came in how I viewed myself. Instead of focusing on what needed to be "fixed" or improved, I began to appreciate qualities I'd previously overlooked:

- The sensitivity I once saw as weakness became something to be grateful for—it allowed me to understand others deeply.
- My perfectionist tendencies, while challenging, showed me my commitment to excellence.
- Even my struggles with self-doubt revealed my capacity for growth and self-reflection.

WHEN GRATITUDE FEELS HARD

There will be times when finding gratitude feels impossible—when life presents challenges that seem to have no silver lining or when you're deep in complex emotions. I encountered this often in my journey, and here's what I learned. When gratitude feels forced or inauthentic, it's often because

- we're trying to bypass legitimate emotions that need to be felt.
- a situation feels too raw or painful.
- we're struggling to accept what is.
- our old patterns of thinking are strongly activated.

As Alex Korb explains in his book *The Upward Spiral*, by consciously practicing gratitude we can train our brain to attend selectively to positive emotions and thoughts, thus reducing anxiety and feelings of apprehension. So, instead of pushing yourself to feel grateful when you don't, try this approach:

1. **Acknowledge where you are:** *I'm finding it hard to feel grateful right now, and that's okay.*
2. **Allow space for all emotions:** Remember what we learned about emotional release in Chapter 4. Sometimes, we need to feel our challenging emotions fully before gratitude can naturally arise.
3. **Start tiny:** When gratitude feels impossible, begin with the most basic things—the breath in your lungs, the ground beneath your feet, or the fact that you're aware enough to notice your resistance.

☆ DEVELOPING YOUR GRATITUDE PRACTICE

The power of gratitude comes through consistent practice. While spontaneous moments of appreciation are beautiful, intentionally cultivating gratitude creates lasting transformation. This practice goes beyond traditional gratitude lists to work with whatever wants to emerge in your life. Here's how I developed what I came to think of as my gratitude lens through journaling.

STEP 1: CREATE A SACRED SPACE

Set aside quiet time with your journal when you won't be interrupted. This isn't about rushing to list three things you're grateful for—it's about diving deeper.

STEP 2: FIND YOUR ENTRY POINT

There are several ways to begin your journaling session:

- **Work with what's present:** Write about an emotion, situation, or pattern you're noticing in your life.
- **Explore resonant ideas:** Reflect on something you've read that has touched you deeply, exploring how it connects to your past or present experiences.
- **Listen to your heart:** Close your eyes, bring your awareness to your heart center, and wait for the first word that arises—it might be a feeling like "openness" or something from your environment like "the ocean."
- **Follow what's alive:** Write about whatever feels most energetically charged in your life right now.

STEP 3: EXPLORE FULLY

Allow yourself to express yourself honestly about whatever has emerged. Don't rush to find the positive—let yourself explore all aspects of the experience or insight.

STEP 4: FIND THE HIDDEN GIFTS

This is where the transformation happens. Ask yourself the following questions:

- What is this teaching me?
- How is this helping me grow?
- What strengths am I developing through this?
- What insights am I gaining?
- How does this connect to my larger journey?

STEP 5: CLOSE WITH GRATITUDE

End your journal entry by explicitly stating what you're grateful for about this exploration. Remember, gratitude can be found in the insights gained, the awareness developed, or even simply in having the space to explore.

☆ PRACTICAL TOOL: PRESENT MOMENT GRATITUDE

While journaling offers a space for deep reflection, practicing gratitude in the present moment creates another powerful dimension of transformation. This practice can happen anywhere, at any time. Here are three key areas where you can cultivate present-moment gratitude throughout your day.

NATURE CONNECTION

- Pause to notice the details around you—the play of light, the movement of leaves, the feeling of the breeze.
- Engage all of your senses—what can you see, hear, feel, smell?
- Allow yourself to feel appreciation for the simple miracle of being alive in this moment.

BODY AWARENESS

- Notice your breath—the simple, constant gift of life.
- Feel your heart beating—the steady rhythm that keeps you alive.
- Appreciate your body's wisdom in communicating with you through sensations and feelings.

DAILY MOMENTS

Transform routine experiences into opportunities for gratitude:

- that first sip of morning coffee or tea
- the warm water of your shower
- the comfort of your bed at night
- the smile of a stranger
- the convenience of technology

The key is to notice these moments and pause to feel gratitude in your body. Let it fill your heart, even for a few seconds. Over time, these small moments of appreciation will combine with your deeper journaling practice to create a more comprehensive gratitude practice—one that will transform both how you reflect on your experiences and how you move through each day.

CREATING A GRATITUDE-RICH LIFE

By combining these practices—the deep reflection of journaling and present-moment awareness—I discovered that gratitude shifted from something I did to something I lived. This integration happens naturally when you do the following.

LAYER YOUR PRACTICES

- Begin the day with journaling to set an intentional foundation.
- Practice present-moment gratitude throughout the day.
- End the day by appreciating what unfolded.

NOTICE THE RIPPLE EFFECTS

As gratitude becomes more natural, you might notice the following:
- greater resilience when challenges arise
- more ease in finding the gifts in difficult situations
- increased capacity to appreciate yourself and others
- a natural tendency to focus on growth rather than criticism
- more energy as you spend less time in lower-frequency emotions

WORK WITH RESISTANCE

Sometimes, you'll encounter moments when gratitude feels impossible or inauthentic. These are opportunities to deepen your practice:
- Acknowledge when gratitude feels hard.
- Be curious about what's blocking appreciation.
- Remember that you can feel challenging emotions and still find something to be grateful for.
- Use your journaling practice to explore these moments of resistance.

The goal isn't to force gratitude or bypass difficult emotions. Instead, you're developing the capacity to hold both—to acknowledge challenges while remaining open to the gifts they might contain.

BUILDING A FOUNDATION FOR SELF-LOVE THROUGH GRATITUDE

The practice of gratitude transforms not just how we see ourselves but how we experience our entire world. As we cultivate an appreciation for everything around us—the natural world, the people in our lives, daily moments of beauty, and even life's challenges—we naturally develop a deeper capacity for self-love.

Through my journey, I witnessed an important truth: When we open our hearts in gratitude to all of life, this expanded appreciation naturally flows back to ourselves. Through this practice

- gratitude for nature reminds us that we, too, are part of life's magnificent unfolding.
- appreciating others helps us recognize these same qualities within ourselves.
- being thankful for our daily experiences opens us to the gift of our own presence.
- finding gratitude in challenges reveals our innate resilience and capacity for growth.

What begins as a conscious practice gradually becomes a natural state of being. We start to

- see ourselves as part of the greater web of life for which we're grateful.
- recognize our interconnection with everything we appreciate.
- experience self-love as a natural extension of our gratitude for life itself.
- meet ourselves with the same appreciation we extend to the world around us.

Remember: Self-love grows most naturally when we cultivate gratitude for the entire tapestry of life—including ourselves as an essential thread within it. Through these practices, gratitude becomes more than just an exercise; it becomes a way of being that embraces all of life, creating a foundation for the deepest form of self-love

> ### 📌 KEY TAKEAWAYS: GRATITUDE AS A GATEWAY TO SELF-LOVE
>
> - Gratitude is more than positive thinking—it's a practice that rewires your brain and transforms your relationship with yourself and the world.
> - The practice of gratitude can encompass past experiences, present moments, and future visions, creating a rich tapestry of appreciation.
> - Journaling with the intention of finding gratitude, even for your challenges, helps transform your perception of difficulties into opportunities for growth.
> - Present-moment gratitude—whether in nature, daily activities, or relationships—strengthens your capacity for self-love and appreciation.
> - Resistance to gratitude is normal and can become an opportunity for deeper understanding and growth.
> - Gratitude for life's full spectrum—from daily moments to intense challenges—naturally expands your capacity for self-love.
> - By regularly practicing gratitude, it shifts from something you do to a lens through which you view yourself and your world.
>
> **Remember:** Each moment of gratitude, whether for yourself or the world around you, opens your heart a little wider to receiving and experiencing love.

Taking a stand
and setting healthy boundaries
*is one of the most
loving things*
you can do —
not only for yourself

but for the other person
as well.

LOUISE HAY

CHAPTER 7:

Setting Boundaries as an Act of Self-Love

Like many aspects of my self-love journey, my relationship with boundaries reflected my journey of learning to honor and value myself fully. While I could maintain boundaries in simpler situations, something shifted when I was faced with potential pushbacks or disagreements. In these moments, my tendency to compromise my needs would surface—not primarily out of fear but from a deeper pattern of not fully honoring my worth.

Looking back, I can see how this selective boundary challenge reflected where I still needed to deepen my self-love. In professional settings, I would take on more work than I could reasonably handle, stay silent when treated inappropriately, or go along with requests even when they didn't align with my capacity or values. In friendships, I consistently put others' needs before my own, offering support and time even when it left me depleted and unable to tend to my own well-being. In romantic relationships, this pattern manifested as accepting behavior that didn't honor my worth, making excuses for actions that crossed my boundaries rather than addressing them directly.

This wasn't just about avoiding conflict. Through my study of neuroscience, I understood that my brain had developed patterns of prioritizing others' needs above my own. When situations called for standing firm in my worth and honoring my needs, I would often default to compromise rather than risk the tension of maintaining my boundaries, especially if doing so might create discomfort.

The journey to maintaining healthy boundaries in these challenging moments paralleled my deepening relationship with self-love. As I began to value and honor myself truly, I discovered something profound: Expressing my needs and maintaining boundaries, especially in situations where it felt challenging, wasn't selfish—it was an essential expression of self-love and self-respect. Esther Perel, a renowned psychotherapist and best-selling author, captures this beautifully in her 2017 book *The State of Affairs: Rethinking Infidelity*.

> *"Boundaries are the distance at which I can love you and me simultaneously."*
> **–Esther Perel**

The first few times I held firm in honoring my needs despite external pressure, it felt uncomfortable. Yet with each boundary I maintained, each time I chose to honor my worth despite the urge to compromise, something remarkable happened. Not only did I grow more confident in expressing my needs, but I was actively rewiring my brain to recognize and honor my worth. What had once felt like an insurmountable challenge began to feel natural—even essential to my well-being.

UNDERSTANDING BOUNDARIES THROUGH A NEUROSCIENCE LENS

Through my exploration of neuroplasticity, I discovered that our ability to set and maintain boundaries is deeply connected to our neural pathways. As we discussed in Chapter 2, boundary-setting behaviors are patterns that can be transformed through conscious awareness and practice.

Research has shown that physical and social pain share neural pathways in our brains (Eisenberger, 2012). This helps explain why boundary violations can feel physically uncomfortable—they trigger the same neural circuits that process physical pain. When someone crosses our boundaries, our brain processes it as a genuine threat to our well-being, which is why we might experience physical sensations and emotional distress.

When we repeatedly compromise our boundaries, we strengthen the neural pathways that prioritize others' needs over our own. This isn't just a behavioral pattern—it's a physical reality in our brain's structure. Each time we silence our needs or accept behavior that doesn't honor our worth, we reinforce these neural pathways, making it more likely that we'll repeat this pattern in the future.

However, the power of neuroplasticity means we can create new, healthier patterns. Each time we consciously choose to honor our boundaries, we begin forming new neural pathways. These new pathways connect boundary-setting with positive outcomes, gradually replacing the old associations of discomfort or unworthiness.

Understanding this neuroscientific perspective helped me understand why setting boundaries initially felt so challenging. It wasn't just about making different choices—I was rewiring years of neural programming that had connected boundary-setting with discomfort and unworthiness. Each time I honored my boundaries, I wasn't just changing my behavior; I was physically restructuring my brain's pathways to align with greater self-love and self-respect.

RECOGNIZING WHEN BOUNDARIES ARE NEEDED

Our bodies often signal the need for boundaries before our conscious mind catches up. These physical and emotional responses aren't random—they're our system's way of communicating that our boundaries are being crossed. Some key signs I learned to recognize included

- ➢ **PHYSICAL SIGNALS**
    - » a tightness in the chest or stomach when agreeing to something
    - » feeling drained after interactions with certain people
    - » tension headaches or muscle tension
    - » disrupted sleep patterns when dealing with boundary issues

- ➢ **EMOTIONAL SIGNALS**
    - » a persistent sense of resentment
    - » feeling taken advantage of
    - » anger that seems disproportionate to the situation
    - » guilt about taking time for yourself
    - » a nagging feeling that your needs aren't being met

The process of recognizing when boundaries are needed involves strengthening the connection between your conscious awareness and your body's wisdom in a process of learning to trust your signals. Here's how I learned to develop this awareness:

- **Regular check-ins:** I began practicing regular moments of awareness throughout my day, especially during or after interactions with others. I would ask myself the following questions:
  - How does my body feel right now?
  - What emotions are present?
  - Am I feeling energized or depleted?
  - Are my needs being honored in this situation?
- **Journaling practice:** I started documenting situations where I felt my boundaries had been crossed, noting both the circumstances and my responses. This helped me identify patterns and triggers that needed attention.
- **Energy awareness:** I learned to monitor my energy levels as a key indicator. When I noticed myself feeling consistently depleted after certain interactions or commitments, it was often a sign that stronger boundaries were needed.

Consistently practicing this awareness made me trust my inner signals more readily. What had once felt like confusing physical or emotional responses became clear messages guiding me toward better self-care and stronger boundaries.

★ PRACTICAL STEPS FOR SETTING BOUNDARIES

Once you recognize the need for boundaries, the next step is learning how to set and maintain them effectively. In my journey, I discovered that setting boundaries wasn't simply about saying "no"—it was about honoring my worth and creating space for genuine self-love to flourish. I learned that effective boundary-setting requires clear, compassionate communication. Instead of apologizing for having needs or making excuses, I began to express my boundaries directly while maintaining kindness:

- Instead of saying, "I'm so sorry, but I probably can't help with that project…" I learned to say, "I appreciate you thinking of me, but my schedule is full, and I need to honor my existing commitments."
- Instead of saying, "Maybe we could talk about this another time…" I learned to say, "I need some time to process this. Can we discuss it tomorrow when I can give it my full attention?"

This shift in communication was about more than just words—it was about rewiring my brain's response to setting boundaries. Each time I communicated clearly and directly, I strengthened new neural pathways that connected boundary-setting with self-respect rather than guilt or fear.

Like any new skill, setting boundaries becomes easier with practice. I began with situations where the stakes felt lower:

- declining additional work projects when my plate was full
- setting limits on my availability for nonurgent communications
- expressing my preferences in everyday situations

As these smaller boundaries became more natural, I was better equipped to maintain my boundaries in more challenging situations. This gradual approach allowed my brain to build new, positive associations with boundary-setting, making standing firm in more difficult circumstances progressively easier.

HANDLING PUSHBACK

One of the most challenging aspects of setting boundaries is managing others' responses. Some people might

- express disappointment or try to make you feel guilty.
- attempt to negotiate or push past your stated limits.
- react with anger or defensiveness.
- try to minimize your needs.

Remember: These reactions often say more about the other person's boundary issues than about your right to set limits. When faced with pushback, I learned to

- remain calm and centered on my worth.
- restate my boundary clearly if needed.
- resist the urge to overexplain, justify, or get caught up in looping conversations.
- trust that my needs are valid, even if others disagree.

CREATING SUSTAINABLE CHANGE

Maintaining boundaries isn't about being rigid or inflexible—it's about creating sustainable patterns that honor yourself and your relationships. I discovered that healthy boundaries often lead to

- more authentic connections.
- increased energy and creativity.
- greater self-respect.

- a deeper capacity for genuine giving.
- an enhanced ability to be present for others.

BOUNDARIES IN PROFESSIONAL SETTINGS

My journey with boundaries began in the workplace when I realized how often I compromised my well-being to be seen as capable and cooperative. I learned that professional boundaries aren't just about saying no to extra work—they're about establishing a framework that allows you to thrive while maintaining your integrity.

Some key professional boundaries I developed included

- being clear about my working hours and availability.
- communicating directly when the project scope began to expand.
- addressing inappropriate behavior or comments promptly.
- setting realistic timelines rather than agreeing to impossible deadlines.
- taking my breaks and vacation time without guilt.

BOUNDARIES IN PERSONAL RELATIONSHIPS

Personal relationships often present the most challenging boundary situations because they involve deeper emotional connections. Through my journey, I discovered that strong boundaries create the foundation for more authentic and fulfilling relationships.

Key areas where I learned to establish clear boundaries included

- emotional energy and support.
- time and availability.
- personal space and privacy.
- financial matters.
- physical and intimate boundaries.

For example, I learned to say things like, "I care about you, but I'm not in a place to offer support right now. Can we talk tomorrow when I can be more present?" or "I need some alone time to recharge. This isn't about you—it's about honoring what I need to stay balanced."

BOUNDARIES WITH FAMILY

Family boundaries can be particularly challenging because they often involve long-established patterns and emotional dynamics. However, my journey with family boundaries taught me that it's possible to maintain loving relationships while still honoring your own needs.

Some essential family boundaries I developed are as follows:

- setting limits on unsolicited advice
- maintaining privacy about personal decisions
- creating space between triggering conversations and responses
- establishing holiday and celebration boundaries
- managing expectations around time and attention

DIGITAL BOUNDARIES

In our constantly connected world, digital boundaries have become increasingly important for my well-being. Setting clear limits around technology and accessibility help me maintain my energy and focus.

The digital boundary practices I implemented include

- defining specific times for checking emails and messages.
- setting expectations about response times.
- using "do not disturb" features during focus time.
- creating boundaries around social media use.
- maintaining my privacy in online spaces.

PHYSICAL SPACE AND ENERGY BOUNDARIES

Learning to honor my physical space and energy needs was crucial for maintaining my overall well-being. This included

- creating a dedicated workspace that others respect.
- maintaining personal space in shared environments.
- setting boundaries around lending personal items.
- protecting my rest and recovery time.
- honoring my body's signals about physical limits.

Implementing these boundaries across different areas of my life, I discovered that each context required slightly different approaches. However, I always returned to the same core principle: honoring my worth and well-being while maintaining respect for others.

★ PRACTICAL EXERCISES

THE BOUNDARY AUDIT

This exercise helps you assess your current boundary landscape and identify areas needing attention.

Take out your journal and create three columns:
- situation/relationship
- current boundaries
- desired boundaries

Spend 15–20 minutes reflecting on different areas of your life:
- work relationships
- personal relationships
- family dynamics
- digital interactions
- physical space and energy

For each area, note the following:
- Where do you feel drained?
- When do you feel resentful?
- Where do you wish you could say no?
- What situations leave you feeling compromised?

THE BODY SCAN BOUNDARY CHECK

This exercise strengthens your ability to recognize your body's boundary signals:

1. Find a quiet space and sit comfortably.
2. Close your eyes and take three deep breaths.
3. Think of a recent situation where your boundaries were crossed.
4. Notice:
    » Where do you feel tension in your body?
    » What sensations arise?
    » How does your breathing change?
    » What emotions surface?
5. Write down these physical and emotional responses.

Practice this exercise regularly to build trust in your body's wisdom about boundaries.

BOUNDARY SCRIPT-WRITING

Create a collection of boundary statements that feel authentic to you. Practice saying them aloud to build confidence.

Start with: "I appreciate the invitation, but…", "I need to honor my commitments to…", "That doesn't work for me because…", and "I value our relationship, and I need…"

Write three versions of each one:
1. Gentle but firm
2. Direct and clear
3. Kind but non-negotiable

THE ENERGY BANK ACCOUNT

This exercise helps you track how different interactions affect your energy levels.

Create a simple journal entry each day, listing the following:

- **Energy deposits:** Activities and interactions that energized you
- **Energy withdrawals:** Situations that drained you
- **Balance:** How you feel at the day's end

Use this information to

- identify patterns that drain your energy.
- recognize where your boundaries need strengthening.
- plan your day to maintain your energy balance.

BOUNDARY VISUALIZATION

This exercise helps rewire your brain's response to boundary-setting:

1. Find a quiet space.
2. Close your eyes and breathe deeply.
3. Visualize yourself
    » standing firm in your worth.
    » communicating your boundaries clearly.
    » maintaining calm when faced with pushback.
    » feeling strong and centered in your choices.
4. Notice how this confident version of you

- » holds their body.
- » speaks their truth.
- » maintains inner peace.

5. Practice embodying these qualities in your daily life.

Complete each exercise in your journal, noting the insights and patterns that emerge. Remember, these practices aren't just about changing behavior—they're about rewiring your brain's relationship with boundaries and strengthening the foundation of your self-love.

COMMON CHALLENGES AND HOW TO NAVIGATE THEM

Setting and maintaining boundaries often presents challenges. During my journey, I encountered several common obstacles that initially felt overwhelming but eventually became opportunities for deeper growth in self-love.

One of my most persistent challenges was managing the guilt that arose when setting boundaries. This guilt often manifested as thoughts like

- ➤ Am I being selfish?
- ➤ Maybe I should just help them anyway.
- ➤ What if they need me?
- ➤ Am I being too rigid?

I learned that this guilt was often connected to old neural pathways that equated self-care with selfishness. Each time I chose to maintain my boundaries despite the guilt, I created new neural connections that aligned with self-respect and worth. In his 2012 book *Letting Go: The Pathway of Surrender*, David R. Hawkins emphasizes that guilt often arises when we set boundaries, but this feeling stems from lower vibrational states.

> "Guilt is a lower vibration emotion that clouds your ability to act in alignment with your highest self. Letting go of guilt frees you to choose what truly serves your well-being."
> –David R. Hawkins, *Letting Go: The Pathway of Surrender*

People's reactions to our boundaries can also be challenging to navigate. Some common responses I encountered included

- emotional manipulation.
- passive-aggressive behavior.
- direct confrontation.
- silent treatment.
- attempts to make me feel guilty.

Preparing for these reactions helped me stay grounded in my boundary decisions. Remember: It's not your responsibility to manage others' reactions to your boundaries.

There will be times when you find yourself slipping back into old patterns. I experienced this particularly during

- high-stress periods.
- family gatherings.
- holiday seasons.
- times when I had pressing work deadlines.
- periods of emotional vulnerability.

Instead of seeing these moments as failures, I learned to view them as opportunities for deeper awareness and practice. Each time we notice ourselves reverting to old patterns, we can choose differently and strengthen our new, healthier pathways.

MAINTAINING BOUNDARIES OVER TIME

One challenge that surprised me was the ongoing nature of boundary work. I initially thought the work was done once I set a boundary. However, I learned that boundaries often need

- regular reassessment.
- occasional adjustments.
- consistent reinforcement.
- clear communication.
- ongoing commitment.

This isn't about failure—it's about growing and adapting as we and our relationships evolve.

The journey of boundary-setting, like all aspects of self-love, is ongoing. Be patient with yourself as you develop these new skills and patterns. Each small step forward is progress, and each boundary honored is an act of self-love.

📌 KEY TAKEAWAYS:
BUILDING AND MAINTAINING HEALTHY BOUNDARIES

- Setting boundaries is a vital expression of self-love, showing both yourself and others that you honor your worth and well-being.
- Your brain can be rewired through neuroplasticity to create and maintain healthier boundary patterns. Each

time you honor your boundaries, you strengthen these new neural pathways.
- Physical and emotional signals serve as important guides to your boundary needs. Learning to recognize and trust these signals is key to developing strong boundaries.
- Clear, compassionate communication is essential for effectively setting and maintaining boundaries. It's possible to express your needs both kindly and firmly.
- Starting with smaller boundaries and gradually building to more challenging ones allows you to develop confidence and trust in your boundary-setting abilities.
- Others' reactions to your boundaries reflect their own patterns, and it's not your responsibility to manage them. Standing firm in your worth is more important than managing other people's responses.
- Regular boundary check-ins and adjustments help ensure your boundaries continue to serve your growth and well-being as you evolve.
- Consistency and practice are essential—each time you maintain a boundary, you strengthen the foundation of your self-love.

Remember: Setting and maintaining healthy boundaries isn't about building walls; it's about creating a sacred space where you and your relationships can thrive. In this space, self-love flourishes, authentic connections deepen, and your true self can fully emerge.

I asked my body,
"What do you need from me?"
It replied softly,
"To be heard, to be honored,
to be loved."

For the first time, I listened.

FIONA SOUTTER

CHAPTER 8:

Physical Self-Love—The Gentle Awakening

The most life-changing transformations often happen when we're focused elsewhere. As I immersed myself in understanding energy and frequency, something unexpected began to unfold: My relationship with my physical self started to shift. Not through force or discipline, but through a natural awakening to my body's needs.

For years, I had approached physical health the way many of us do—as something to be conquered. Exercise meant pushing myself through strenuous workouts at the gym or forcing myself to run, activities that seemed impossibly daunting from where I stood. So instead, I did very little, caught in the belief that it was too overwhelming.

As I mentioned in an earlier chapter, my body had been trying to tell me something was wrong for years. Migraines were my constant companion, striking at the slightest hint of stress or emotional disturbance. These weren't just headaches; they were my body's desperate attempts to communicate what my younger self couldn't process—the impact of childhood trauma and accumulated stress. Yet, like so many signals from my body, I had learned to push through them, wearing them almost as a badge of honor or proof of my resilience.

The shift began subtly. As I focused on raising my emotional frequency, I naturally found myself drawn to practices that supported my physical well-being—grounding exercises, time in nature, drinking better-quality water, early-morning light exposure, gentler lighting in the evenings. These weren't part of a prescribed health regimen; they were simply what felt right as I became more attuned to my body's needs.

I still remember the morning when the magnitude of this transformation hit me. I woke up excited for my morning routine—a walk on the beach then a swim in the ocean, followed by reading, journaling, and meditation. Even during winter, when the ocean was bracingly cold, it didn't feel like a challenge to be overcome. These weren't exercises to be endured; they were moments to be savored. What struck me most was how natural it all felt, and how far removed this was from my previous "no pain, no gain" mentality.

The most surprising part? As these gentle practices became part of my daily life, I found myself naturally wanting to do more. The energy that came from honoring my body's rhythms led to better food choices and, eventually, to more vigorous forms of exercise. But this time, it wasn't coming from a place of "should" or "must"—it was coming from a place of genuine care for my physical self.

This was physical self-love in its truest form—not a regimen imposed from outside but a natural outgrowth of deeper self-awareness. The practices that made the biggest difference weren't the ones that pushed me to my limits but the ones that allowed me to restore and reconnect: grounding exercises, permission to rest without guilt, hot–cold therapy. These seemingly simple activities were revolutionary because they challenged my core belief that stopping meant I wasn't proving my worth to the external world.

UNDERSTANDING THE BODY'S LANGUAGE

What I came to realize was that our bodies speak to us constantly, but we've been conditioned to ignore or override these messages. The migraines I experienced weren't just random occurrences—they were manifestations of deeper patterns, physical expressions of emotional wounds that hadn't been processed. When we're operating at a low frequency of self-love, we often interpret these signals as inconveniences to be pushed through rather than messages to be heeded.

This dismissal of our body's wisdom shows up in countless ways. We ignore fatigue because we "have to" finish that project. We suppress hunger or eat mindlessly because we're "too busy" to have a proper meal. We dismiss aches and tensions because they don't fit into our schedule. Each time we override these signals, we're telling our body that its needs don't matter—a profound expression of low self-love.

THE FREQUENCY OF PHYSICAL SELF-LOVE

Just as our thoughts and emotions operate at different frequencies, our physical self-care practices emit their own energetic signatures. What I discovered was that gentle, restorative practices often carried a higher frequency than the punishing routines I'd previously associated with "getting healthy." This wasn't just about the physical actions themselves—it was about the energy and intention behind them.

When we approach physical health from a place of self-punishment or rigid discipline, we're operating at a frequency of fear and control. This might produce short-term results, but it's unsustainable because it's misaligned with our body's natural wisdom. In contrast, when we approach our physical well-being from a place of self-love and respect, we tap into a higher frequency that makes healthy choices feel natural and effortless.

This understanding transformed how I viewed restorative practices:

- Grounding wasn't just about connecting with the Earth; it was about allowing my body to reset its natural rhythms.
- Rest wasn't a sign of weakness but a crucial part of building strength.
- Hot–cold therapy wasn't about enduring discomfort but about awakening my body's innate healing capabilities.
- Time in nature wasn't just pleasant—it was essential for recalibrating my entire system.

THE MIND-BODY CONNECTION: BEYOND TRADITIONAL UNDERSTANDING

> *"Your body temple is the only place your Spirit has to live; care for your temple, and it will give you wings."*
> —Dr. Espen Wold-Jensen

While these gentle physical practices were transforming my health, I began to understand something even more significant: The healing wasn't coming just from the practices themselves but from the dramatic shift in my thoughts, beliefs, and emotional patterns. This understanding deepened when I discovered the work of Dr. Bruce Lipton, a pioneering cell biologist and former faculty member of the University of Wisconsin's School of Medicine. His research in epigenetics—the study of how environmental signals, including our thoughts and beliefs, control our genes—provided the scientific framework that explained my healing.

Through my work with Dr. Espen on releasing stored emotional trauma and transforming limiting beliefs, my chronic migraines had completely disappeared. This wasn't just symptom management—it was as if I had switched off the genetic expression that had manifested as migraines for most of my life. But how was this possible?

Lipton's research reveals something revolutionary: Our genes are not our destiny. As he explains (Gustafson, 2017):

> The new science called epigenetic control sounds almost like the same thing [as genetic control]. When I say genetic control, it translates as "controlled by genes." The new science is called epigenetics. It sounds similar, but it is profoundly different. Epi means "above," so when I say epigenetic control, I am literally saying, "control above the genes." This is the new biology. It reveals that the environment and our perception of the environment are what control our genetic activity.

This understanding transformed how I viewed my healing journey. My migraines weren't just random occurrences or a genetic predisposition I had to live with. They were physical manifestations of stored trauma and limiting beliefs, particularly my deep-seated belief that I was "a migraine sufferer." By changing my internal environment through emotional release and belief transformation, I had literally changed the signals being sent to my cells.

Modern epigenetic research confirms that emotional trauma creates lasting changes in gene expression through various mechanisms, including DNA methylation (Nie et al., 2022; Jiang et al., 2019). However, these changes aren't necessarily permanent. According to Lipton in his 2005 book *The Biology of Belief*, our perceptions and beliefs can influence our biology at the cellular level through chemical signaling, suggesting potential pathways for healing the effects of trauma on gene expression. When we address stored trauma and limiting beliefs, we're not just doing psychological work—we're creating tangible changes at the cellular level, as Lipton's research demonstrates.

This realization helped explain why the gentle, frequency-raising practices I was naturally drawn to were so effective. Each time I chose rest over pushing through, each moment I spent grounding on the beach or swimming in the ocean, I wasn't just performing physical activities—I was sending new signals to my cells. I was creating an internal environment of safety, nurturing, and self-love, allowing my body's natural healing capabilities to activate.

THE POWER OF EARTH CONNECTION

Understanding this mind–body connection through epigenetics helped me appreciate why grounding—one of the first practices I was naturally drawn to—had such a huge impact on my health. What began as enjoyable morning walks on the beach was actually a powerful reset for my entire biological system.

Through my studies with quantum biology expert Dr. Catherine Clinton, I developed an understanding of why these Earth-connecting practices are so powerful. When we directly connect with the Earth's surface, we tap into a natural flow of energy that affects our cellular function. This connection goes beyond simple relaxation—it influences our body at the quantum level, affecting everything from blood flow to nervous system function.

The science supporting these benefits is compelling. A 2013 study published in *Integrative Medicine* showed that just two hours of earthing increased the zeta potential (the negative charge on red blood cells) by an average of 270%. This greater charge helps the cells repel each other, reducing clumping and improving blood flow. The researchers concluded that earthing "reduces blood viscosity and clumping" and "appears to be one of the simplest and yet most profound interventions for helping reduce cardiovascular risk and cardiovascular events" (Menigoz et al., 2020).

The same research demonstrated, through speckle contrast laser imaging, that earthing generates rapid improvements in facial blood flow and enhances the autonomic nervous system's regulation of peripheral circulation (Menigoz et al., 2020). This improved circulation explains why I often felt more energized and clearer after my morning beach walks—my entire body was receiving better blood flow and oxygenation.

THE INTELLIGENCE OF WATER

My morning beach practice wasn't just about grounding. It also included another powerful element of healing: water.

The ocean swims that began as intuitive choices led me to a deeper understanding of water's role in our health, through both immersion and consumption.

OCEAN IMMERSION

Those early-morning ocean swims, particularly during winter, weren't just invigorating—they were therapeutic on multiple levels. Clinton's analysis of the research shows that exposure to the negative ions found in ocean spray increases mitochondrial function, creating a cascade of biological effects that improves brain, immune, and nervous system function (Clinton, 2024). This aligns with broader scientific findings that demonstrate how negative air ions can influence multiple aspects of human physiology. A comprehensive review of the research evidence shows that the presence of negative air ions is associated with improvements in psychological health and overall well-being, with studies documenting their effects on various biological processes including blood flow, inflammatory responses, and stress reduction (Jiang et al., 2018).

Beyond the benefits of negative ions, the cold exposure from ocean swimming triggers its own powerful healing responses in my body—something I'd experienced intuitively but has now been validated by science. Regular cold water immersion offers multiple evidence-backed benefits (Esperland et al., 2022). Research shows that cold exposure strengthens our immune system, helping our bodies fight illness more effectively (Shevchuk & Radoja, 2007). When we combine the storytelling and the science, it all makes sense. What cold water swimmers have experienced anecdotally, researchers are now confirming in their studies: The practice enhances our physical health and mental well-being, from boosting blood flow to lifting our mood. The evidence keeps growing, supporting what I'd felt all along during those early-morning ocean swims.

THE LIVING MATRIX OF WATER

As my journey of self-love deepened, I found myself naturally wanting to honor my body in every way possible. This led me to explore the teachings of various holistic health experts, and what I discovered about water shocked me. Despite believing I was taking care of myself by drinking plenty of water, I learned that the quality of my water—both tap and bottled—was potentially causing more harm than good.

Through Clinton's work, I discovered that water plays an even more fundamental role in our health than most realize. We are essentially water beings—70% water by weight and almost 99% water by molecular count. What's more, the water we drink begins entering our bloodstream within five minutes, making its quality crucial for our immediate and long-term health.

While many of us are fortunate to have access to clean, treated drinking water—something many others in the world still lack—recent research has revealed some concerning findings about both tap and bottled water. One study found that many water systems contain various contaminants, including arsenic, uranium, nitrates, and manmade chemicals known as PFAS, which can persist in the environment for decades (Haederle, 2023). Municipal water treatment, while necessary for public safety, often requires multiple chemical processes that can affect water's natural properties.

Bottled water presents its own challenges. Ground-breaking research found that a single liter of bottled water contains an average of 240,000 plastic particles, with about 90% being nanoplastics—particles small enough to enter the body's cells and tissues (Contie, 2024). Scientists have already detected these plastic particles in human blood, lungs, gut, and other tissues, though the long-term health implications are still being studied.

This understanding deepened through my studies with David Avocado Wolfe, a renowned expert in health, eco, nutrition, and longevity. His insights into water quality helped me understand why not all water is created equal. Wolfe advocates for "living water"—naturally alkaline, mineral-rich water sourced straight from natural springs, free from plastic contamination and processing.

The power of spring water lies in its journey through the Earth's natural filtration system. As it moves through sometimes hundreds of thousands of feet of filtration material, it becomes enriched with essential minerals like silica, magnesium, and calcium, while also gathering healthy microbes and probiotics. This process creates something far more complex than any artificial filtration system could achieve.

Perhaps most fascinating is the structured nature of spring water. As it emerges from the Earth, its molecules are arranged in a cohesive hexagonal form—a structure that makes it more biologically available to our cells. This structured arrangement also gives water its remarkable ability to carry information. Research documented in Carly Nuday's 2014 book *Water Codes* reveals that small clusters of water molecules can contain hundreds of thousands of information panels, leading physicist Rustum Roy to describe water as potentially "the single most malleable computer."

Just as eating processed food differs from consuming whole, living food, drinking treated or bottled water differs fundamentally from consuming pure, structured spring water. This living water carries not just hydration but the intelligence and vitality of the Earth itself.

So, after learning these truths about water, it was as if I couldn't "unsee" them. I began viewing water differently, and it became a priority to make conscious choices about the water I drank wherever possible. While I understood it might not always be feasible, I committed to consuming "living water" as much as I could, recognizing its vital role in my body's health and vitality. This wasn't about perfectionism—it was about honoring my body's need for pure, life-giving nourishment, another expression of deepening self-love.

THE POWER OF NATURAL LIGHT

While the ocean and sand were powerful healing elements of my morning routine, another crucial factor was at play: early-morning sunlight. Just as I'd discovered the importance of natural, living water, I began to understand how natural light—particularly morning light—was essential for our biological functioning.

Those early-morning beach visits weren't just peaceful—they were resetting my entire biological clock. Research reveals that humans, like almost all living organisms, have evolved with biological clocks synchronized to the Sun's 24-hour cycle (Foster, 2021). Morning light advances our circadian rhythm, while evening light delays it. Our modern lifestyle, particularly exposure to artificial light at night and electronic devices, has disrupted this vital connection (Meléndez-Fernández et al., 2023). The timing, intensity, duration, and wavelength of light all play crucial roles in maintaining our biological rhythms (Duffy & Czeisler, 2009), as we are beings primed to respond to the Sun's natural cycles.

The science behind this is fascinating. Morning sunlight exposure helps regulate our circadian rhythm, the internal clock that governs nearly every biological process in our body. This natural rhythm influences

- hormone production, especially cortisol and melatonin.
- metabolic function.
- immune system response.
- mental clarity and mood.
- sleep quality.

Perhaps most surprisingly, I learned that proper light exposure could even influence inflammation and histamine levels in the body. The mast cells that release histamine are governed by their own circadian clock. When we live in misalignment with the solar rhythm, we create dysfunction within these cells, potentially increasing inflammation and stress responses.

The contrast between natural sunlight and artificial lighting became increasingly apparent to me. While modern LED lighting might be energy efficient, it often emits high levels of blue light and electromagnetic fields that can disturb our natural rhythms. Just as I'd learned to be discerning about water quality, I began to understand the importance of light quality—both getting enough natural light during the day and protecting myself from artificial light, especially in the evening hours.

MOVEMENT AS MEDICINE

The transformation in how I approached movement was perhaps one of the greatest shifts in my journey. As mentioned earlier, I had spent years caught between two extremes: either pushing myself through punishing workouts or doing nothing at all, paralyzed by the belief that exercise had to be strenuous to be effective. This all-or-nothing mentality had left me feeling overwhelmed and disconnected from my body's natural desire to move.

The change began with those morning beach walks. There was no pressure, no performance metrics to meet—just the simple pleasure of moving my body through nature. Walking barefoot on the sand, feeling the texture beneath my feet, the ocean breeze on my skin... it became a form of moving meditation. Swimming, too, transformed from an exercise into a dance with the ocean. Even in winter's cold embrace, these movements weren't about burning calories or building muscle—they were about connecting with myself and my environment.

What I didn't realize then was how these gentle movements were supporting my body's natural healing systems. Research shows that fluid, natural movement is crucial for lymphatic flow—our body's cellular cleaning system (Tatlici & Çakmakçi, 2021). Unlike our blood, which is pumped by the heart, the lymphatic system relies entirely on movement and muscle contraction to circulate. In fact, this research found that even minimal movement makes a difference—just a short walk of eight steps doubled the flow rate in lymphatic vessels. The gentle, rhythmic movements I was naturally drawn to were actually ideal for supporting this essential detoxification process, as the lymphatic system depends on both muscle contractions and changes in fluid pressure to maintain circulation.

The most surprising part, as I mentioned earlier in the chapter, was how this gentle approach led to me wanting more. As my energy increased and my body felt more alive, I found myself naturally drawn to other forms of movement. It wasn't driven by "should" or "must," but by genuine desire. My body, no longer afraid of being pushed beyond its limits, began to trust that movement could be enjoyable and restorative rather than depleting.

This wasn't just about physical movement—it was about moving energy through my body in a way that felt nurturing rather than punishing. Whether it was a morning swim, a mindful walk, or, eventually, more vigorous activities, each movement became an expression of self-love rather than self-improvement.

THE SACRED ROLE OF REST

Perhaps the most countercultural aspect of my physical self-love journey was learning to honor rest. In our achievement-oriented society, rest is often viewed as laziness or weakness—a narrative

I'd deeply internalized. Just as I'd previously believed exercise needed to be punishing to be effective, I'd convinced myself that pushing through exhaustion was a sign of strength and dedication.

But as my relationship with my body deepened, I began to understand rest differently. This wasn't about collapsing from exhaustion after pushing too hard but about intentionally creating space for restoration. Through Dr. Clinton's work, I learned that rest isn't just about giving our muscles a break—it's essential for cellular repair, nervous system regulation, and even genetic expression.

When we're constantly operating in a state of "go, go, go," our bodies remain in sympathetic nervous system dominance—the fight-or-flight state. This chronic stress response affects everything from our hormone production to our immune function. Regular periods of true rest, on the other hand, activate our parasympathetic nervous system—the rest-and-digest state—allowing our bodies to

- repair and regenerate at the cellular level.
- process and eliminate toxins.
- balance hormone production.
- consolidate memories and learning.
- strengthen immune function.

Understanding this science helped me recognize that rest wasn't a luxury or a sign of weakness; it was a biological necessity. Just as important as my morning beach routine was the practice of honoring my body's need for downtime, whether through adequate sleep, mindful breaks throughout the day, a digital detox, or periods of deeper restoration.

CREATING SACRED SLEEP SPACE

Just as morning light exposure sets our circadian rhythm for the day, evening light plays a crucial role in preparing our bodies for rest. Our ancestors lived by the Sun's natural rhythm, experiencing only warm, reddish light from fire after sunset. Today's artificial lighting, particularly the blue light from screens and LED bulbs, disrupts this natural pattern.

Understanding this, I began creating an evening environment that honored my body's natural rhythms. Several hours before bedtime, I would switch to warmer lighting sources—candles, salt lamps, and halogen bulbs with red tones. This wasn't just about ambience; it was about signaling to my brain that it was time to begin the transition to sleep.

If I needed to use screens in the evening, I would wear blue-light-blocking glasses. But more importantly, I began treating the hour before sleep as sacred time, free from screens and filled instead with gentle, restful activities that helped my nervous system downshift into a parasympathetic state.

This attention to evening light became as important as my morning light exposure. Together, they created a natural rhythm that supported deep, restorative sleep—the kind of rest that allows for true healing and regeneration.

☆ PRACTICAL ACTIVITY: BODY AWARENESS AUDIT

This journaling exercise is designed to help you gently explore your current relationship with your body and identify small, manageable steps toward physical self-love. Remember, there's no pressure to change everything at once. The goal is to listen to your body and honor what feels natural and doable for you right now.

PART 1: CURRENT PATTERNS

Take a moment to reflect on and write about the following:
- How do you currently speak to your body? Notice the tone and words you use.
- What signals from your body do you tend to ignore (hunger, fatigue, tension, etc.)?
- What activities make your body feel good and energized?
- What activities leave you feeling depleted?
- Where do you feel the most pressure or "shoulds" around your physical health?

PART 2: GENTLE EXPLORATION

Consider these questions with curiosity and compassion:
- What small change feels most appealing to you right now?
- What's one way you could make your environment more supportive of rest?
- Is there a form of movement you've always wanted to try?

- What does your body naturally gravitate toward?
- What's one signal from your body you'd like to start honoring more?

PART 3: SMALL STEPS FORWARD

From your reflections, choose *one* small step that feels genuinely appealing and doable. It might be

- getting a salt lamp or candles for evening relaxation.
- taking a 15-minute walk outside.
- drinking more water.
- starting a simple stretching routine.
- creating a calmer bedtime environment.

Remember: There's no need to implement everything at once. Choose what resonates most with you right now. Trust that as you begin honoring your body in small ways, your capacity for self-love will naturally expand.

WEEKLY CHECK-IN QUESTIONS

- How does this small change feel in my body?
- What am I noticing about my energy levels?
- What's working well?
- What might need adjustment?
- What other gentle changes am I feeling drawn to?

This is about nurturing a loving relationship with your body, not about adding more pressure or "shoulds" to your life. Let your journey unfold naturally, one small step at a time.

CREATING YOUR OWN PATH

I've shared my journey of physical self-love—how living by the ocean naturally led to morning swims, how understanding frequency drew me to explore water quality, and how raising my vibration led to gentler ways of moving and resting. But your path will likely look very different from mine, and that's exactly as it should be.

Physical self-love isn't about following someone else's routine or checking boxes on a wellness to-do list. It's about tuning in to what naturally calls to you, what makes your body feel nourished and alive. The practices that emerged in my life were simply a reflection of my environment, my circumstances, and what resonated with my growing self-awareness.

This transformation wasn't driven by external standards or comparisons. There was no target weight to reach, no ideal body to emulate, no competition to win. Instead, it was a natural extension of deepening self-love: As I learned to honor my spirit and mind, honoring my body followed naturally. The increased energy, clarity, and vitality were welcome side effects of this loving attention, not goals I was striving to achieve.

Trust that as you deepen your own journey of self-love, your body will guide you toward what it needs. This might be gentle movement in your backyard, quiet moments of rest, dancing with wild abandon in your living room, hiking in nature, joining a sports team, or any other practice that brings you genuine joy. Some might find peace in stillness, while others discover their body's wisdom through more vigorous expression—there's no right or wrong way. Let your practice emerge naturally, in its own time, in its own way. This is your unique journey of physical self-love, and it will be as individual as you are.

📌 KEY TAKEAWAYS: PHYSICAL SELF-LOVE—THE GENTLE AWAKENING

- The most profound transformations often happen when you're focused elsewhere—physical self-love can emerge naturally as you raise your frequency and deepen your self-awareness,

- Your thoughts and beliefs directly influence your physical health through epigenetics, as demonstrated by how changing our internal environment can affect our gene expression.

- The quality of what you consume—from water to light exposure—impacts you at the cellular level, making conscious choices an act of self-love.

- Gentle, restorative practices often carry a higher frequency than punishing routines, leading to more sustainable transformation.

- Movement becomes medicine when it emerges from joy and genuine desire rather than external pressure or a "should" mentality.

- Rest isn't a luxury but a biological necessity—honoring your need for restoration is essential for transformation.

- Each person's journey to physical self-love is unique—what works for one may not work for another, and that's perfectly natural.

Remember: Physical self-love isn't about following someone else's routine or checking boxes on a wellness to-do list—it's about tuning into what naturally calls to you and honoring your body's wisdom in ways that feel authentic and sustainable for you.

The willingness to show up changes us.

It makes us a little braver each time

BRENÉ BROWN

CHAPTER 9:

Making Self-Love Sustainable—From Practice to Way of Being

I remember the morning, as I watched the first hints of pink and gold streak across the pre-dawn sky, my feet sinking into the cool sand, when I realized something extraordinary had shifted in my journey. What had begun years ago as a collection of separate practices—meditation, journaling, emotional release work, boundary-setting—had transformed into something far more integrated. Self-love was no longer something I did; it had become who I was.

Standing on the beach, I reflected on how different this moment felt compared to my early days of trying to piece together a path to self-love. Back then, each practice felt like a separate task, something to check off a list or remember to do. These elements now flowed together naturally, showing up in ways I hadn't planned or forced.

My morning beach routine illustrated this transformation perfectly. What had started as a conscious decision to create more mindful mornings had evolved into an organic rhythm that nourished my soul. I had learned to trust my inner guidance rather than forcing myself to follow a rigid schedule.

But this morning ritual was just one visible sign of a deeper transformation. The real shift had happened in how I moved through all aspects of my life. The tools and practices we've explored throughout this book—from observing thought patterns to maintaining boundaries, from practicing gratitude to releasing emotional blocks—had been woven into my daily experience. They were no longer separate practices but different expressions of the same fundamental truth: I was worthy of love, starting with my own.

This chapter is about how that integration happened. It's about how various pieces of the self-love journey—all the tools, practices, and insights we've explored—come together to create lasting change. More than that, it's about how what begins as conscious practice eventually transforms into a natural way of being. Ultimately, sustainable self-love isn't about perfecting a set of practices—it's about becoming someone who naturally expresses love toward themselves in every aspect of life.

IMMERSING IN WISDOM

As this integration occurred in my daily practices, I made another crucial choice that accelerated my transformation: I became a dedicated student of personal growth. This wasn't just about reading an occasional self-help book or attending the odd workshop. It was about consciously restructuring my daily input—choosing to feed my mind with content that supported my evolution rather than reinforcing old patterns.

My morning beach walks became a time for listening to personal development podcasts that expanded my understanding of self-love and consciousness. When driving in my car, I chose podcasts or audiobooks from teachers whose work resonated with my journey instead of music or news. My evening unwinding time shifted from mindless TV to watching talks and interviews with thought leaders on YouTube or Gaia TV.

This wasn't about constantly forcing myself to consume "serious" content. Rather, it was about becoming intentional about what influenced my thinking. I began to see my mind as a garden—one where I could choose what to plant and nurture. Why feed it content that reinforced self-doubt or negative patterns when I could nourish it with wisdom that supported my growth?

Social media, once a source of comparison and self-criticism, transformed into a tool for inspiration. I carefully curated my Instagram feed to follow teachers, authors, and individuals who shared authentic insights about personal growth and health. Instead of scrolling through endless posts that left me feeling lacking, I created a feed that regularly reminded me of the truths I was integrating into my life.

What fascinated me was how this immersion naturally deepened my understanding of the practices we've explored in this book. Something I heard in a podcast might shed new light on a concept from Dispenza's work. A YouTube video might offer a fresh perspective on emotional release. An Instagram post might remind me about boundary-setting on a challenging day.

This constant exposure to growth-oriented content served multiple purposes:

- It reinforced the new neural pathways I was building.
- It provided fresh insights and perspectives on familiar concepts.
- It reminded me that I wasn't alone on this journey.
- It offered practical tools and strategies I could implement.
- It kept me inspired and motivated, especially during challenging times.

Most importantly, this immersion in wisdom helped normalize these new ways of thinking and being. When you consistently expose yourself to ideas about self-love, emotional intelligence, and personal growth, these concepts gradually become your default way of seeing the world. The more I surrounded myself with these teachings, the more natural it felt to think and act from a place of self-love.

This wasn't about becoming dependent on external wisdom—quite the opposite. It was about creating an environment that supported my growth until these new ways of thinking and being became fully integrated. Like learning any new language, immersion accelerates fluency. I was learning the language of self-love, and surrounding myself with it helped me become fluent faster.

THE NATURAL FLOW OF INTEGRATION

As I immersed myself in wisdom and practice, the tools and concepts I was learning started showing up naturally in my daily life, often in unexpected ways. What had begun as conscious practices became intuitive responses to life's moments, both challenging and ordinary.

This integration showed up in countless small ways throughout my day:

- speaking with quiet confidence in meetings where I once would have diminished my contributions
- tuning into my body's wisdom when facing decisions rather than second-guessing things endlessly
- creating space to feel and process difficult emotions instead of suppressing them
- setting boundaries without guilt or overexplanation
- celebrating achievements without minimizing them
- taking breaks when my body signaled the need without justification
- speaking about myself with kindness in casual conversations

The organic nature of this integration made it sustainable. Rather than forcing myself to follow rigid practices, I learned to flow with what each moment called for. The wisdom I consumed through podcasts, books, and videos wasn't just accumulating as knowledge—it was actively reshaping how I moved through my days. When a friend offered a compliment, the teachings about receiving with grace would naturally arise. During stressful projects, the understanding of maintaining high-frequency emotional states would guide my responses.

I understood that sustainable self-love isn't about executing a set of practices perfectly. It's about creating an internal and external environment where loving yourself becomes the natural response. Just as a plant will naturally grow toward the Sun when given the right conditions, self-love naturally expresses itself when we create the right conditions in our lives.

This transformation manifested across several key areas: my morning routine, which set the foundation for each day: how I navigated life transitions and adapted these practices during change, how I handled challenging moments, and how I engaged with myself and others in daily life. Let's explore each of these areas to understand how theory became practice and practice became a way of being.

THE FOUNDATION: SACRED MORNING TIME

The most precious way in which this transformation manifested in my daily life was in my morning routine. I discovered that how I started my day—before the world could make demands on my time and energy—profoundly influenced everything that followed.

My mornings began before dawn when the world was still quiet. This wasn't about forcing myself to be an early riser—it was about creating space that was truly mine, free from external pressures and interruptions. Whether it was my beach walks or simply sitting in my garden with a cup of tea, this time belonged to me alone.

What made this morning so powerful wasn't the specific activities—though I cherished my beach walks, ocean swims, and quiet meditation moments—but the quality of presence I could bring to it. Without emails to check, calls to return, or obligations to meet, I could fully drop into being with myself.

Starting each day by consciously choosing activities that elevated my emotional state created a foundation that carried through my entire day. When challenges arose later, I could more easily return to that centered state I'd established in the morning. It was like filling my emotional tank before the day made demands on it.

The impact of this sacred morning time extended far beyond the actual hours I spent in practice. I was making a powerful statement about my worth by prioritizing this time for myself and putting my self-care first in my day. This wasn't selfish; it was strategic. By filling my cup first, I had more to offer others throughout the day.

NAVIGATING LIFE TRANSITIONS

One of the biggest tests of sustainable self-love comes during major life transitions. While my experience with expanding my business required adaptation, life transitions come in many forms—each demanding its own kind of flexibility and self-awareness.

MAJOR LIFE TRANSITIONS

These pivotal moments might include

- becoming a new parent.
- starting or ending significant relationships.
- career changes or job losses.
- moving to a new city or country.
- health challenges.
- the loss of a loved one.
- beginning or ending important chapters in life.

During these transitions, it's easy for self-love practices to fall away just when we need them most. The key isn't maintaining everything exactly as it was—that's often impossible and can create unnecessary pressure. Instead, it's about pausing to consciously adapt our practices to our new reality.

ADAPTATION STRATEGY

When facing major transitions, I learned to follow these steps:

1. **Pause and reflect**

    » Take stock of what's changing in your daily rhythm.
    » Identify which practices feel most essential for your well-being.
    » Consider what's realistically possible in your new circumstances.

2. **Creative adaptation**

    For example, if becoming a new parent means your hour-long morning routine is no longer feasible, you might do the following:

    » Break your practice into smaller segments throughout the day.
    » Find ways to integrate self-love moments while caring for your baby.
    » Create new rituals that honor both your needs and your new role.

3. **Release perfectionism**

    » Understand that adaptation isn't compromise—it's wisdom.
    » Let go of how things "should" look.

» Trust that even modified practices can be powerful.

REAL-LIFE EXAMPLES

New Parenthood

When your sleep patterns are disrupted and time feels scarce

- transform diaper changes into moments of presence and connection.
- use feeding times for meditation or breathing practices.
- find ways to include your child in your self-care rituals.

Relationship Changes

Whether you're beginning or ending a relationship

- use the transition as an opportunity to strengthen your self-trust.
- maintain boundaries around your essential practices.
- allow your routines to evolve while keeping their core purpose.

Career Transitions

When your professional life shifts dramatically

- review and adjust the timing of your self-love practices.
- find ways to integrate self-love into your new schedule.
- use the change as an opportunity to establish new, supportive habits.

The key is remembering that these practices aren't rigid rules to follow but rather tools for supporting ourselves. During major transitions, they might need to shift and change, but their essence—the commitment to self-love—remains constant.

Life transitions often trigger our old patterns of self-neglect. This is precisely when maintaining some form of self-love practice becomes most crucial. Even if it looks different from before, keeping this connection to ourselves provides stability and support during times of change.

MAKING PRACTICES PORTABLE: MY JOURNEY WITH TRAVEL

While major life changes like becoming a parent or changing careers require significant adaptation, even shorter-term transitions like regular travel can test our commitment to self-love practices. When my business expanded and suddenly required frequent travel, I faced my own challenge: how to maintain the sacred morning routine that had become my foundation when I was no longer waking up in the same environment each day.

What I discovered was both humbling and empowering. Rather than abandoning my practices or becoming rigid about replicating them exactly, I learned to capture their essence in ways that could travel with me. My experience offers one example of how we can adapt while maintaining our commitment to self-love.

TRAVEL ADAPTATIONS

My sacred morning routine became portable through a few essential elements:

- ➤ My journal and a meaningful book became non-negotiable travel companions.

- I used my phone mindfully for guided breathwork or meditation.
- I rose at dawn to explore peaceful spots near my accommodation—whether a quiet park or a serene urban corner.
- I created sanctuary spaces during travel with noise-canceling headphones for meditation during flights.
- I found moments of stillness before the world awakened.

SCHEDULE SHIFTS

Being in new places taught me to honor my joy alongside my responsibilities:

- I allowed myself to explore and enjoy new surroundings without guilt.
- I arranged my work hours flexibly—sometimes working in the early morning, late afternoon, or evening—to accommodate both joy and productivity.
- I trusted that taking time to nourish my soul actually enhanced my work rather than detracted from it.
- I recognized that self-love meant giving myself permission to experience life fully, not just to work efficiently.

This experience taught me something valuable about sustainable self-love: It's not about maintaining rigid practices but about staying true to the essence of what nourishes us, even as the form of what that is changes.

THE LANGUAGE OF SELF-LOVE

Perhaps the most comprehensive transformation in my journey was how self-love became woven into my way of speaking—to myself and others. This wasn't a conscious practice of affirmations or forced positive self-talk. Rather, just as my morning rituals had evolved from a conscious effort to a natural rhythm, my internal dialogue shifted from critical to compassionate without a forced effort.

What started as catching and correcting negative self-talk became a natural flow of supportive internal conversation. When facing challenges, words of encouragement would arise naturally: *You've got this*, or *Take your time; you're learning*. This wasn't about suppressing real feelings or maintaining artificial positivity—it was about speaking to myself with the same kindness I would offer a dear friend.

This shift wasn't instant; it evolved gradually through consistent awareness and practice. The neural pathways we discussed earlier—the ones we can strengthen through conscious choices—had created new default patterns in my internal dialogue. What once required constant vigilance now flowed naturally, like a stream finding its path down a mountainside.

This transformation showed up in countless daily moments:

- during business meetings where I once would have diminished my contributions
- in social situations where I previously might have self-deprecated
- when looking in the mirror and noticing my reflection

- while celebrating achievements, both big and small
- in moments of challenge or perceived failure

This new way of speaking to myself created a major ripple effect. The more I naturally spoke to myself with compassion and understanding, the more authentic my interactions with others became. I no longer felt the need to diminish myself to make others comfortable or to overexplain my choices. This wasn't just about changing words. Instead, it was about embodying a deeper truth: that I was worthy of the same kindness, patience, and respect I so readily offered others.

This internal shift also meant that when old patterns of self-critical talk did surface—because they sometimes still do—I could notice them with gentle awareness rather than judgment. Instead of seeing these moments as failures in my self-love practice, they became opportunities to choose again, to remember who I had become, and to return to the language of self-love that now felt like my native tongue.

CREATING A CIRCLE OF SUPPORT

The transformation in how I spoke to myself naturally extended to how I cultivated relationships. I discovered that sustainable self-love thrives best when supported by conscious connections. This wasn't about dramatically changing my entire social circle—it was about being intentional about the energy I allowed into my space and how I showed up in relationships.

One of the most beautiful aspects of this transformation was how it naturally influenced my friendships. I had shared my journey with close friends, being open about the changes I was making and the patterns I was working to shift. This vulnerability, rather than pushing people away, created deeper connections. When I would momentarily slip into old habits—perhaps diminishing an achievement or speaking self-critically—these friends would gently reflect back to me: "I noticed you downplaying that accomplishment," or "Remember how far you've come."

This supportive awareness extended beyond just catching moments of self-criticism. I remember a particularly powerful moment during a conversation with a friend and business colleague. We were discussing business growth, and I made a comment that revealed a limiting belief about money. With genuine kindness, she gently brought my attention to how that belief might be holding me back. Rather than feeling judged or defensive, I felt immense gratitude for her insight. It was a beautiful example of how the right kind of support can help us spot the subtle ways in which old patterns show up—even when we think we're being confident and clear. That moment allowed me to pause, recognize the limiting belief, and consciously choose a new perspective that aligned with my growth.

These interactions became precious mirrors, reflecting not just obvious moments of self-doubt but also the more subtle ways in which our old programming can surface in casual conversation. Having friends who could compassionately point out these moments—whether they were self-critical comments, limiting beliefs, or unconscious self-sabotage—became an invaluable part of my growth. It wasn't about policing each other's words; it was about creating a space where we could all grow more conscious of the stories we tell ourselves and choose more empowering narratives.

This openness about my journey had an unexpected ripple effect. By showing up authentically and sharing my struggles and discoveries, I created spaces where others felt safe to explore their own relationship with self-love. Conversations naturally deepened beyond surface-level interactions to meaningful exchanges about growth, challenges, and transformation.

THE EVOLUTION OF RELATIONSHIPS

- Some friendships naturally deepened as they aligned with my growth.
- Others gently shifted or faded as our paths diverged.
- New connections emerged with people for whom with this way of being resonated.
- Existing relationships transformed as I showed up more authentically.

CREATING SUPPORTIVE AGREEMENTS

I learned to have open conversations with close friends about how we could support each other's growth. This might sound like one or more of the following:

- "I'm working on receiving compliments gracefully. Would you be willing to notice when I deflect praise?"
- "I'm practicing setting boundaries. Can we agree to be honest with each other about our needs?"
- "I'm learning to celebrate my achievements. Can we create space to acknowledge our wins together?"

These weren't formal agreements but shared understandings that helped us grow together. The key was maintaining these connections from a place of self-love rather than one of dependency. I wasn't looking for others to validate my worth—I was creating relationships that reflected and reinforced the love I was cultivating for myself.

This shift in how I approached relationships became another indicator of how deeply self-love had transformed my way of being. I no longer felt compelled to maintain connections that drained my energy or required me to shrink myself. Instead, I naturally gravitated toward relationships that supported mutual growth, authenticity, and joy.

THE POWER OF MICRO-MOMENTS

While my morning rituals and supportive relationships created a strong foundation for sustainable self-love, I discovered that the real magic happened in what I came to call "micro-moments"—those small, seemingly insignificant choices that appeared throughout my day. These weren't grand gestures or lengthy practices but rather brief moments of choosing self-love that, when strung together, created lasting transformation.

These micro-moments showed up in countless ways:

- **In business settings**
  - taking a conscious breath before responding to a challenging email
  - pausing to celebrate small wins instead of rushing to the next task

- » speaking up in meetings with confidence rather than self-doubt
- » setting realistic deadlines instead of overcommitting

- **In personal care**
  - » that moment of appreciation while applying moisturizer
  - » choosing nourishing food because my body deserves care
  - » taking a few deep breaths between tasks
  - » going to bed when I felt tired rather than pushing through

- **In relationships**
  - » receiving compliments with a simple "thank you" rather than deflecting
  - » honoring my need for alone time without guilt
  - » expressing my preferences clearly and kindly
  - » setting boundaries in the moment rather than after resentment built

- **In my internal dialogue**
  - » catching self-critical thoughts and choosing compassion instead
  - » acknowledging my growth with genuine appreciation
  - » meeting mistakes with curiosity rather than judgment
  - » allowing myself to feel difficult emotions without rushing to fix them

What made these micro-moments so powerful was their accessibility. I didn't need special time, tools, or circumstances to practice them. They could happen anywhere—in a grocery store line, during a work meeting, or while stuck in traffic. Each small choice to honor myself built upon the others, creating a tapestry of self-love woven throughout my day.

Moreover, these moments became natural pauses in my day—brief opportunities to check in with myself and choose consciously rather than react habitually. While I didn't always make the most self-loving choice (because this journey isn't about perfection), having an awareness of these moments meant I could more often choose responses that aligned with my worth.

Over time, I noticed something remarkable: These micro-moments began to multiply naturally. One conscious choice led to another, creating a momentum of self-love that carried me through my days. What started as occasional moments of awareness became a natural way of moving through life—not because I was forcing it but because choosing self-love had become my default state.

This is where true sustainability in self-love practices emerges—not in the big gestures or formal practices (though these are important) but in these small, consistent choices to honor ourselves throughout our days. Each micro-moment becomes both a practice of self-love and evidence of how far we've come in our journey. Figure 9.1 illustrates how these practices come together to form a framework to support self-love.

Figure 9.1: The neuroscience of self-love—a complete framework

The Neuroscience of Self-Love: A Complete Framework

Awareness	Integration	Expression
Daily Practices	**Core Processes**	**Living Practices**
• Observer Perspective	• Neuroplasticity Exercises	• Boundary Setting
• Thought Pattern Recognition	• Emotional Release Work	• Gratitude Practice
• Emotional Awareness	• Pattern Interruption	• Authentic Communicaiton
• Body Scanning	• Future Self Embodiment	• Self-Compassion
Neural Impact	**Neural Impact**	**Neural Impact**
• Strengthened Pre-frontal Cortex	• New Neural Pathways	• Positive Neural Networks
• Enhanced Neural Monitoring	• Reduced Stree Response	• Enhanced Emotional Regulation
• Increased Self-Awareness	• Improved Neural Integration	• Strengthened Self-Reference
Outcomes	**Outcomes**	**Outcomes**
• Reduced Reactive Patterns	• Emotional Freedom	• Healthy Relationships
• Clear Inner Guidance	• Behavioral Flexibility	• Natural Self-Love
• Emotional Intelligence	• Sustainable Change	• Life Balance

Neural Foundations of Self-Love
Supported by Neuroplasticity, Emotional Intelligence, and Conscious Awareness

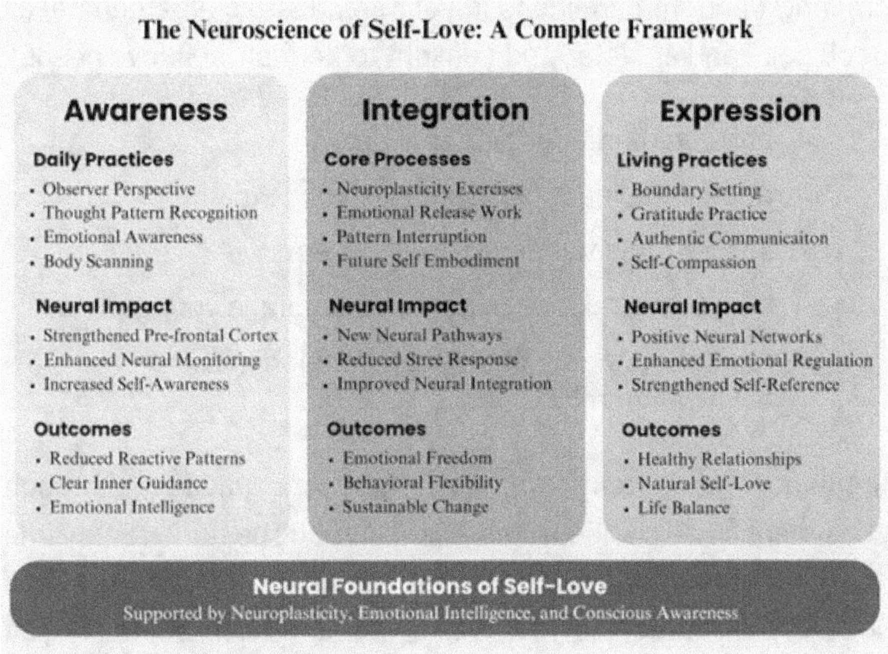

THE JOURNEY TO INTEGRATION

As we conclude this chapter on making self-love sustainable, remember that true transformation isn't about perfectly implementing every practice we've discussed. It's about finding your rhythm and allowing these elements to weave into your life naturally.

Your journey might begin with creating sacred morning time—even if it's just 15 minutes before the rest of your household wakes. Or perhaps you'll start with pattern interrupts, using Dispenza's "change" technique when you notice old thought patterns arising. Maybe you'll begin by bringing awareness to your self-talk, supported by friends who understand your journey.

The key is recognizing that sustainable self-love grows from creating the right conditions in your life. Just as a garden needs good soil, regular water, and sunlight to flourish, self-love needs

- protected time for personal growth.
- nourishing input that supports your evolution.
- simple tools to interrupt old patterns.
- supportive relationships that encourage your growth.
- regular moments of self-compassion and acknowledgment.

Remember that transformation happens gradually, through consistent small choices rather than dramatic gestures. Each time you choose self-compassion over self-criticism, each morning you devote yourself to your well-being, and each boundary you maintain with grace—these moments compound over time, creating lasting change.

Most importantly, trust that as you continue this journey, what begins as conscious practice will naturally evolve into a way of being. Just as I discovered on that quiet beach at dawn, you too will find that self-love becomes less about what you do and more about who you are.

📌 KEY TAKEAWAYS: MAKING SELF-LOVE SUSTAINABLE

- Create conditions that support your growth through protected time and nourishing input.
- Use simple pattern interrupts to create new possibilities in challenging moments.
- Allow transformation to happen gradually through consistent small choices.
- Build supportive relationships that encourage your evolution.
- Trust that conscious practice will naturally evolve into a way of being.
- Remember that sustainable self-love is about integration, not perfection

Remember: Your journey to sustainable self-love is uniquely yours. Let it unfold in its own time, in its own way, knowing that each small step toward self-compassion creates ripples of transformation in your life.

Each small choice
to honor yourself,
each moment
of conscious awareness,
and each boundary maintained
is rewiring your brain
for greater self-love

and creating ripples
of transformation in your life.

FIONA SOUTTER

CONCLUSION:

From Practice to Transformation

As we reach the conclusion of our journey together, I want to share something that happened recently. While preparing for an important business presentation, I noticed a familiar flutter of anxiety in my stomach. But instead of spiraling into self-doubt or pushing the feeling away, I naturally moved through the tools and practices we've explored in this book. I took a moment to observe the sensation with curiosity rather than judgment. I acknowledged the old neural pathway of "not enough" trying to activate and consciously chose a different response. With gentle gratitude, I appreciated how this moment of awareness itself showed how far I'd come.

This small moment crystallized something profound. The practices we've explored in this book aren't just techniques to use when things get tough—they'll become your natural way of being. The neuroscience of self-love isn't about forcing yourself to think positive thoughts or maintain rigid practices. It's about rewiring your brain's fundamental patterns so self-love becomes your default state.

Through understanding how neuroplasticity works, you've learned that your brain can be rewired for self-love. You've discovered how to step back from old thought patterns and create new ones by practicing awareness. Emotional release work has allowed you to begin freeing up the energy you once used to suppress your feelings, while gratitude practice has supported you to start building new neural pathways that support self-appreciation. And finally, through boundary-setting you've learned to honor your worth in tangible ways.

But perhaps most importantly, you've learned that this journey isn't about reaching some mythical destination of perfect self-love. Instead, it's about creating a sustainable foundation for growth and transformation that will continue long after you've finished this book.

I share this now as someone who's walked this path, who's moved from severe anxiety and depression to a place where I can

- maintain healthy boundaries in both my personal and professional relationships.
- navigate challenges with greater ease and resilience.
- trust my inner wisdom rather than seeking constant external validation.
- experience genuine joy and connection in my relationships.
- focus on my physical health without the previous patterns of self-neglect.

This transformation didn't happen overnight, and it wasn't always easy. There were moments of doubt, days when old patterns tried to reassert themselves, and times when maintaining these practices felt challenging. But with each small choice to honor myself, each moment of choosing awareness over automation, each boundary maintained, the new neural pathways grew stronger.

The migraines that once plagued me have subsided. The romantic relationship I'm in reflects the level of self-love I've cultivated. My business relationships have deepened through authentic communication and clear boundaries. While I still encounter challenges and occasionally slip into old patterns, the difference now is that I have the tools to recognize and redirect these moments quickly, preventing them from becoming extended periods of self-doubt or anxiety.

As you continue your journey beyond these pages, remember the following key points:

- Your brain is constantly rewiring itself based on your experiences and choices. Each time you choose self-love over self-criticism, you strengthen those neural pathways.
- Small, consistent choices matter more than grand gestures. The everyday moments of self-respect, boundary-setting, and gratitude create lasting change.
- Progress isn't linear. There will be days when old patterns surface, and that's okay. What matters is your capacity to observe these moments with compassion and choose differently.
- Your journey is unique. While the neuroscience principles we've explored are universal, how you apply them will be distinctive to your life and circumstances.
- Integration happens naturally when you create the right conditions. Just as a garden grows when given proper care, self-love flourishes when you provide the environment it needs.

The tools and practices in this book aren't meant to be another set of "shoulds" in your life. They're invitations to explore what genuine self-love looks like for you. Whether you're facing professional challenges, navigating relationships, or dealing with internal struggles, these practices can adapt to support whatever you're experiencing.

As we close this chapter together, I invite you to see this not as an ending but as a beginning. The neural pathways you've begun creating through these practices will continue to strengthen with use. The awareness you've developed will keep expanding. The boundaries you've learned to set will support ever-deeper levels of self-respect.

You've already taken the most important step by choosing to embark on this self-love journey. Trust that each small choice to honor yourself, each moment of conscious awareness, and each boundary maintained is rewiring your brain for greater self-love and creating ripples of transformation in your life.

Remember: You are worthy of love, starting with your own. And with the tools of neuroscience and the practices we've explored, you have everything you need to continue nurturing that truth into fuller expression in your life.

Your journey to deep, sustainable self-love continues. It's beautiful to witness how each step you take transforms your life and inspires others to begin their own journey of self-discovery and self-love.

Keep going. Keep growing. Keep choosing yourself. The path of self-love awaits, and you are perfectly equipped to walk it.

RESOURCES

Arntz, W., Chasse, B., & Vincente, M. (Directors). (2004). *What the Bleep Do We Know!?* Lord of the Wind, Captured Light.

Beyond Inspiration. (2024, November 23). *Rewire your brain to create a life you love - Joe Dispenza Motivation* [Video]. YouTube. https://www.youtube.com/watch?v=Hm6nEiERhLU

Chopra, D. (2017, June 28). *How to love yourself just as you are.* Deepak Chopra. https://www.deepakchopra.com/articles/how-to-love-yourself-just-as-you-are/

Clinton, C. (n.d.). *Quantum biology of trauma course.* Dr. Catherine Clinton. https://www.drcatherineclinton.com/quantum-biology-of-trauma-course

Clinton, C. [@dr.catherineclinton]. (2024, December 23). Nature heals: Recent research found that negative ions, like those found in rainfall, waterfalls and ocean surf, were associated with [Video]. Instagram. https://www.instagram.com/p/DD7SmAgPYeL/

Contie, V. (2024, January 23). *Plastic particles in bottled water.* National Institutes of Health. https://www.nih.gov/news-events/nih-research-matters/plastic-particles-bottled-water

Dispenza, J. (Director). (n.d.). *Rewired* [13-part series]. Gaia. https://www.gaia.com/series/rewired

Dispenza, J. (2013). *Breaking the habit of being yourself: How to lose your mind and create a new one.* Hay House.

Dispenza, J. (2017). *Becoming supernatural: How common people are doing the uncommon.* Hay House.

Duffy, J. F., & Czeisler, C. A. (2009). Effect of Light on Human Circadian Physiology. *Sleep Medicine Clinics, 4*(2), 165–177. https://doi.org/10.1016/j.jsmc.2009.01.004

Eisenberger, N. I. (2012). The neural bases of social pain: evidence for shared representations with physical pain. *Psychosomatic Medicine, 74*(2), 126–135. https://doi.org/10.1097/PSY.0b013e3182464dd1

Esperland, D., de Weerd, L., & Mercer, J. B. (2022). Health effects of voluntary exposure to cold water – a continuing subject of debate. *International Journal of Circumpolar Health, 81*(1), 2111789. https://doi.org/10.1080/22423982.2022.2111789

Foster, R. (2021). Fundamentals of circadian entrainment by light. *Lighting Research & Technology, 53*(5), 377–393. https://doi.org/10.1177/14771535211014792

Freud, A. (Host). (2023, August 15). *How to regulate your emotions with Dr. Justine Grosso* [Audio podcast episode]. In *Women of Influence*. SheSpeaks. https://shespeaksinc.com/podcast/regulate-your-emotions-dr-justine-grosso/

Fuchs, E., & Flügge, G. (2014). Adult neuroplasticity: More than 40 years of research. *Neural Plasticity, 2014*, 541870. https://doi.org/10.1155/2014/541870

Gustafson, C. (2017). Bruce Lipton, PhD: The jump from cell culture to consciousness. *Integrative Medicine, 16*(6), 44–50. https://pmc.ncbi.nlm.nih.gov/articles/PMC6438088/

Haederle, M. (2023, October 23). *U.S. drinking water often contains toxic contaminants, UNM scientist warns*. The University of New Mexico. https://hsc.unm.edu/news/2023/10/u.s-drinking-water-often-contains-toxic-contaminants-unm-scientist-warns.html

Hawkins, D. R. (1998). *Power versus force: The hidden determinants of human behavior*. Hay House.

Hawkins, D. R. (2014). *Letting go: The pathway of surrender*. Hay House.

HeartMath Institute. (2016). *Science of the heart: Exploring the role of the heart in human performance.* HeartMath Institute. https://www.heartmath.org/research/science-of-the-heart/details/

Hölzel, B. K., Carmody, J., Vangel, M., Congleton, C., Yerramsetti, S. M., Gard, T., & Lazar, S. W. (2011). Mindfulness practice leads to increases in regional brain gray matter density. *Psychiatry Research, 191*(1), 36–43. https://doi.org/10.1016/j.pscychresns.2010.08.006

Howes, L. (Host). (2018, August 13). Heal your body with your mind: Dr. Joe Dispenza [Audio podcast episode]. In *The School of Greatness.* Greatness Media. https://lewishowes.com/podcast/heal-your-body-with-your-mind-dr-joe-dispenza/

Jiang, S. Y., Ma, A., & Ramachandran, S. (2018). Negative air ions and their effects on human health and air quality improvement. *International Journal of Molecular Sciences, 19*(10), 2966. https://doi.org/10.3390/ijms19102966

Jiang, S., Postovit, L., Cattaneo, A., Binder, E. B., & Aitchison, K. J. (2019). Epigenetic modifications in stress response genes associated with childhood trauma. *Frontiers in Psychiatry, 10.* https://doi.org/10.3389/fpsyt.2019.00808

Korb, A. (2015). *The upward spiral: Using neuroscience to reverse the course of depression, one small change at a time.* New Harbinger Publications.

Lipton, B. H., PhD. (Host). (n.d.-a). Controlling genetic expression [Video podcast episode]. In *Inner Evolution with Bruce Lipton.* Gaia. https://www.gaia.com/video/controlling-genetic-expression

Lipton, B. H. (Host). (n.d.-b). Empowered genetics [Video podcast episode]. In *Inner Evolution with Bruce Lipton.* Gaia. https://www.gaia.com/video/empowered-genetics

Lipton, B. H. (2015). *The biology of belief: Unleashing the power of consciousness, matter, and miracles* (10th anniversary edition). Hay House.

Meléndez-Fernández, O. H., Liu, J. A., & Nelson, R. J. (2023). Circadian rhythms disrupted by light at night and mistimed food intake alter hormonal rhythms and metabolism. *International Journal of Molecular Sciences, 24*(4), 3392. https://doi.org/10.3390/ijms24043392

Menigoz, W., Latz, T. T., Ely, R. A., Kamei, C., Melvin, G., & Sinatra, D. (2020). Integrative and lifestyle medicine strategies should include earthing (grounding): Review of research evidence and clinical observations. *Explore, 16*(3), 152-160. https://doi.org/10.1016/j.explore.2019.10.005

Morales, J. I. (2020, November 29). *The heart's electromagnetic field is your superpower: Training heart–brain coherence.* Psychology Today. https://www.psychologytoday.com/au/blog/building-the-habit-of-hero/202011/the-hearts-electromagnetic-field-is-your-superpower

Nie, Y., Wen, L., Song, J., Wang, N., Huang, L., Gao, L., & Qu, M. (2022). Emerging trends in epigenetic and childhood trauma: Bibliometrics and visual analysis. *Frontiers in Psychiatry, 13*. https://doi.org/10.3389/fpsyt.2022.925273

Nuday, C. (2014). *Water codes: The science of health, consciousness, and enlightenment.* Water Ink Publishing.

Roebuck, P. (2021, April 7). *Neuroscience: Have scientists found the inner child?* Paul Roebuck. https://paulroebuck.co.uk/blog/inner-child-in-amygdala

Shetty, J. (2023, October 27). *Jay Shetty & Dr. Joe Dispenza on the consequences of stress and overthinking.* Jay Shetty. https://www.jayshetty.me/blog/jay-shetty-dr-joe-dispenza-on-the-consequences-of-stress-and-overthinking

Shevchuk, N. A., & Radoja, S. (2007). Possible stimulation of anti-tumor immunity using repeated cold stress: a hypothesis. *infectious Agents and Cancer, 2*, 20. https://doi.org/10.1186/1750-9378-2-20

Siegel, C. (2023, December 5). *Deeper into the quantum with Dr. Espen* [Video]. YouTube. https://www.youtube.com/watch?v=2w-c3GssfNo

Singer, M. A. (2007). *The untethered soul: The journey beyond yourself*. New Harbinger Publications.

Singer, M. A. (2015). *The surrender experiment: My journey into life's perfection*. Hachette.

Sorrells, S. F. Paredes, M. F., Velmeshev, D., Herranz-Pérez, V., Sandoval, K., Mayer, S., Chang, E. F., Insausti, R., Kriegstein, A. R., Rubenstein, J. L., Garcia-Verdugo, J. M., Huang, E. J., & Alvarez-Buylla, A. (2019). Immature excitatory neurons develop during adolescence in the human amygdala. *Nature Communications, 10*, 2748. https://doi.org/10.1038/s41467-019-10765-1

Sturgeon, J. A., & Zautra, A. J. (2016). Social pain and physical pain: Shared paths to resilience. *Pain Management, 6*(1), 63–74. https://doi.org/10.2217/pmt.15.56

Tang, R., Friston, K. J., & Tang, Y. Y. (2020). Brief mindfulness meditation induces gray matter changes in a brain hub. *Neural Plasticity, 2020*, 8830005. https://doi.org/10.1155/2020/8830005

Tatlici, A., & Çakmakçi, O. (2021). Exercise and lymphatic system. *Turkish Journal of Sport and Exercise, 23*(2), 150-154. https://dergipark.org.tr/en/pub/tsed/issue/64815/957914

Wold-Jensen, E. (n.d.). *Learn to consciously master every area of your life*. Dr Espen. https://drespen.com/

Wold-Jensen, E. (2019, May 2). *Is this blocking your energy centres and keeping you sick and stuck right now?* [Video]. YouTube. https://www.youtube.com/watch?v=ztma1oWwL6E

Wong, Y. J., Owen, J., Gabana, N. T., Brown, J. W., McInnis, S., Toth, P., & Gilman, L. (2018). Does gratitude writing improve the mental health of psychotherapy clients? Evidence from a randomized controlled trial. *Psychotherapy Research, 28*(2), 192–202. https://doi.org/10.1080/10503307.2016.1169332

ABOUT THE AUTHOR

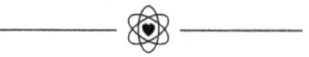

Fiona Soutter's journey to understanding the neuroscience of self-love began unexpectedly. After building a successful digital marketing career, a life-altering car accident in 2014 led her down a path of deep personal exploration. While managing chronic pain and navigating her recovery, she discovered that traditional approaches to healing weren't enough—she needed to understand the deeper connection between mind, brain, and lasting transformation.

This quest led her to an extensive study of quantum physics and neuroscience, particularly through the work of Dr. Joe Dispenza, and a personal mentorship with Dr. Espen Wold-Jensen. What began as a personal journey of healing evolved into a profound understanding of how ancient wisdom and modern neuroscience can work together to create lasting change. Through this work, Fiona discovered how to break free from her self-sabotaging behaviors and transform the limiting beliefs that had held her back.

The impact of this transformation was dramatic. In 2020, armed with this new understanding of herself, Fiona launched an ecommerce business whose turnover grew to seven figures within just 18 months—a success that she attributes not to business strategies alone but also to the inner work that allowed her to step into her full potential.

Today, Fiona combines her entrepreneurial success, her 20 years of teaching experience, and her passion for neuroscience and personal transformation. Drawing on her background as an educator, she excels at making complex concepts accessible and practical. She mentors entrepreneurs in understanding that success comes not from what we do but from who we become.

Fiona is passionate about sharing her journey to help others break free from their own limiting patterns and access their innate capacity for growth and self-love. Through her writing and speaking, she aims to transform the lives of those facing similar challenges—whether they're struggling with self-doubt, healing from trauma, or feeling held back by old programming. Her mission is to bridge the gap between scientific understanding and practical application, making these powerful tools for transformation accessible to everyone.

This book is the culmination of her research, personal experience, and work with clients. By sharing her struggles and breakthroughs, Fiona hopes to illuminate the path for others and show them that lasting transformation is possible when we understand how to work with our brain's natural capacity for change.

You can connect with Fiona at
www.facebook.com/likefionasoutter

MAKE A DIFFERENCE WITH YOUR REVIEW

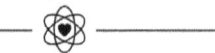

"When you light another's path, you also brighten your own."

People who help others on their journey of self-discovery create ripples of positive change.

Let's create these ripples together! Would you help someone just like you—someone who's ready to transform their relationship with themselves but doesn't know where to start?

My mission is to make the journey to self-love clear and accessible for everyone by bridging science with soul. But to reach more people who need this message, I need your help.

Most people choose books based on reviews. Your honest review could be the lighthouse that guides someone to their own transformation. Your words could help...

…one more person break free from self-doubt

…one more heart heal from past wounds

…one more soul discover their worth

…one more mind understand their power to change

…one more life transform through self-love

To make a difference, simply scan the QR code and leave a review.

Your review might be exactly what someone needs to hear to begin their own journey to self-love. Thank you for being part of this ripple of transformation!

With gratitude,

Fiona

www.ingramcontent.com/pod-product-compliance
Lightning Source LLC
Chambersburg PA
CBHW052141070526
44585CB00017B/1918